GEOLOGY EXPLAINED IN
SOUTH AND EAST DEVON

IN THE SAME SERIES

Geology Explained in the Severn Vale and Cotswolds
by William Dreghorn, BSc, FRGS

Geology Explained in the Forest of Dean and Wye Valley
by William Dreghorn, BSc, FRGS

GEOLOGY EXPLAINED IN SOUTH AND EAST DEVON

by
JOHN W. PERKINS, BA, FRGS

Illustrations by
the Author

DAVID & CHARLES : NEWTON ABBOT

ISBN 0 7153 5304 7

Set in 11pt Cornell
and printed in Great Britain
by Bristol Typesetting Company Limited Bristol
for David & Charles (Publishers) Limited
South Devon House Newton Abbot Devon

Contents

Introduction

To some degree we are all curious about the past. It may be our need to have roots in the district we live in, or to deepen our understanding of places we visit and enjoy. But to read local history alone is not enough, for like our own lives local history, in many respects, was shaped by something older still—by geology.

Historians' questions can often be answered by the geologist. How did these stones come to be here? Why is this hill a particular shape, and why was a certain spot below it chosen as a village site? If a stone-built house needed repairs, where would the same stone be found and what are its special properties? The historian can ask such questions in the Devon landscape today, and he can see how our ancestors answered them before us. Geology was always a living subject for them, even if they studied and obeyed it almost instinctively.

Though now taken for granted—perhaps because we all too rarely stop to think about it—geology influences modern life in countless ways. We depend heavily on the limestones for cement and concrete, on other rocks for road materials. China clay from Dartmoor goes into us, on us and around us, in medicines, paper, china utensils and in dozens of other ways. There would be a serious water shortage without the high granite moorland, whose peat cover acts like an enormous sponge to foster our local rivers. There are geological reasons why the Devon coastline is so richly coloured, so varied in outline and so attractive to the tourist. In short, geology and landscape are alive, living influences as much today as in the past.

As the explorer looks at modern Devon he must remember he sees it at a particular stage in its history—a very unusual stage, too. Only a short time ago geologically it was affected by the great changes of sea level in the Ice Age—a unique event which affected other areas of Britain in similar fashion. But there is one special reason why south-west England has greater importance—the region was never covered by glaciers. In Devon, the geologist can study his rocks unhampered by the mantles of glacial drift which obscure the solid rock outcrops of many other regions. For geologists it is indeed a classical area and there are many other reasons why it remains so.

> Scenery of striking beauty and contrast, fathered by geology of classical repute and bearing a great variety of wild plant and animal life, has attracted geologist, naturalist and tourist interest throughout the years. (Geol. Surv. Mem., Okehampton, 1969)

What is true for Okehampton is true for all Devon. There can hardly be a better-endowed or better-loved county for scenery and geological interest.

It was in the south-west that Geological Survey mapping began in 1820, and in 1839 De La Beche published the results in the Survey's first memoir. Many of the one-inch sheets available today are basically products of the same century, of the days when one had to tether one's horse securely before setting out. They are remarkable maps, produced under difficulties we can hardly envisage and without many of the aids a modern geologist can summon to his help.

Some of the Devon sheets have, of course, been revised since the earlier productions. A fine map and memoir for Okehampton appeared in 1969, the Teignmouth sheet is revised and work progresses into North Devon—for which maps have not been available for some years. The establishment of a district office of the Institute of Geological Sciences in Exeter is testimony to the renewed study of the region.

There are still unsolved problems in Devonshire geology, different points of view and interpretation—part again of the living nature of the subject. Some of the unanswered questions and alternative ideas are given here.

South and East Devon offers something for everyone interested in landscape—whether it is the straightforward enjoyment of its scenery or the discovery of its geological secrets and treasures. There is the fascinating work of the sea on the varied coastal exposures, or the variety of rock ages. You can go back to days of warm Devonian oceans, when fish were first evolving on earth; then to arid deserts, beneath the Cretaceous seas and out of them again, or into the intensely cold Ice Age. Each fascinating episode played its part in creating South Devon.

South Devon's geology is like a gigantic jig-saw puzzle in three dimensions, but the problem is not how to make it but to decipher how it was made. It is a game of detection and we cannot have all the clues we would like—only a cutting here, a borehole there, to give us a glimpse of what lies inside. Fortunately, the rich coastline provides ample natural cross-sections. Quarry sites and block diagrams will help the reader with the more poorly exposed districts between the coast and the moors (Dartmoor and the Tamar Valley will be described in a later volume by the author).

The basic ingredients of the county's rolling landscape are the high moorland centre, the surrounding lowlands bevelled to various heights and deeply trenched by rivers, and the sinuous coastline with its penetrating estuaries and grand cliffs. Written for all who love South Devon, either as a tourist area or a place to live in, this book aims to deepen their understanding and enjoyment. It may also help to popularise geology in a wider sense, and should remind us that we are tenants of a heritage millions of years old, and one that we must do our best to conserve.

It has been assumed throughout that the reader will constantly have a compass, a one-inch geological map and a one-inch Ordnance Survey map at hand. As a help, the grid reference and direction of the view are given on all illustrations where they may be of use. Inevitably, this is sometimes a notional guide since both large trees and buildings have a preference for standing in front of geological items!

CHAPTER 1

Plymouth and District

Physically and historically the greatest of Devonshire harbours, Plymouth has witnessed the coming and going of countless people, famous and otherwise. Its Sound is perhaps one of the best-known stretches of water in the world. The steep slopes of Staddon Heights to the east, the gentle but more famous ones of the Hoe, and many other parts of its shore have served as vantage points for numerous historic occasions.

Always in the role of a national arsenal and defence centre —often to the neglect of its commerce—Plymouth has been fortified at many periods in its history. The military, like the sightseers, were never slow to appreciate the commanding positions offered by the surrounding cliffs. On the slopes of the Hoe, Henry VIII raised artillery towers, and in 1666 Charles II crowned its eastern end with the Citadel, building it in limestone and granite. On the eastern shore, the round Mount Batten tower dates from the same period, while the slopes above are crowned with 'Palmerston's follies' of 1860 vintage—Forts Stamford, Staddon and Watchhouse. With its cliffs providing historic sites and its broad estuaries safe and sheltered harbours, in Plymouth geology and history are intimately related.

The one-inch Geological Survey map reveals one of the notable features of the district, an unbroken sequence from Lower Devonian to Upper in the space of a few miles (Map 1). The lowest visible beds, the Dartmouth Slates, fringe the outermost arms of the Sound at Bovisand and Penlee Point. Inside the line of the breakwater, Meadfoot Beds and Staddon Grits form high, plateau-like ground with steep cliff profiles. The Middle Devonian commences just south of Mount Batten with the slate beds at Jennycliff.

The eastern coast of the Sound is still a problem for

MAP 1

(Generalised from maps of the Institute of Geological Sciences with permission of the Director)

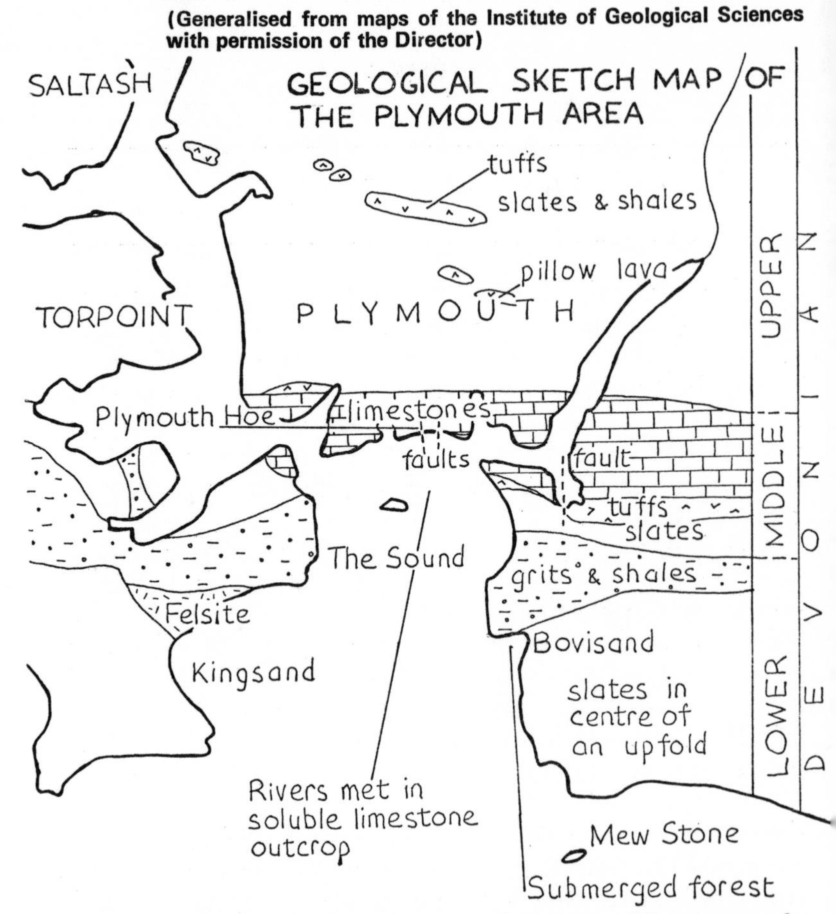

geologists. The old divisions quoted here would not now be valid if a re-survey was made. Many of the grits, for example, are simply variations in the type of deposit rather than separate major units from different periods of Devonian time.

On the north side of the Sound the Middle Devonian limestones outcrop in the Hoe. They cover the whole area of this ridge and, descending Armada Way into the City Centre, reach almost to the line of Notte Street, where they give way to Upper Devonian shales, known locally as 'shillet'. A great thickness of this shale is represented from this point northwards to Tavistock. Most of Plymouth's housing areas and its shopping centre stand on it. It is a rather weak, easily shattered rock when exposed and quarry men disliked it because of the unstable nature of its faces.

A natural exposure of it can be seen in the rock garden of the Drake's Circus roundabout, by the pedestrian tunnels leading to Cobourg Street. The other part of the rockery is man-made, but provides a good idea of the appearance of undressed blocks of local limestone.

Another good shillet exposure occurs on Lipson Hill on the way down to Alexandra Road. Here the typical red colour of the Upper Devonian is seen. Trenches and roadworks in many parts of the city provide further brief reminders of it.

THE SOUND AND ITS ESTUARIES

The story of the Sound is best taken up at the time when the sea had receded from its highest levels of the Ice Age. The role of those higher levels in the history of rivers like the Dart is described in Chapter 5. The Tamar and the Plym were also affected by these events, but with one important difference. When the shoreline had retreated beyond the present coastline, they flowed westwards across the land now lying beneath the Channel. The other South Devon rivers belonged to a separate eastward-flowing system.

With a little imagination the explorer can easily reconstruct the scenery of Plymouth in those days. Modern place names must serve as guides. From Forder Valley, Lipson and Tothill, steep little valleys joined the Plym. Near the Laira Bridge, the Plym entered a limestone gorge. Its sides were steep and cragged, continuous blocks and faces of limestone rising among the grass-covered slopes. Here and there a few trees might find enough water to survive but usually only grass could live on such well-jointed rocks where water quickly drained away underground. Parts of the plateau-like hilltops may have been completely bare, forming natural limestone-pavements comparable to parts of the Pennines today.

As the Plym descended the gorge, it absorbed water trickling from the cave systems in the hillsides. A narrow, and probably very beautiful little canyon would have marked the entry of a small tributary stream flowing from the Radford Valley and Hooe Lake. Beyond this junction the Plym turned westwards to meet the river Tamar.

FIG 1

THE LANDSCAPE OF PLYMOUTH SOUND BEFORE SUBMERGENCE

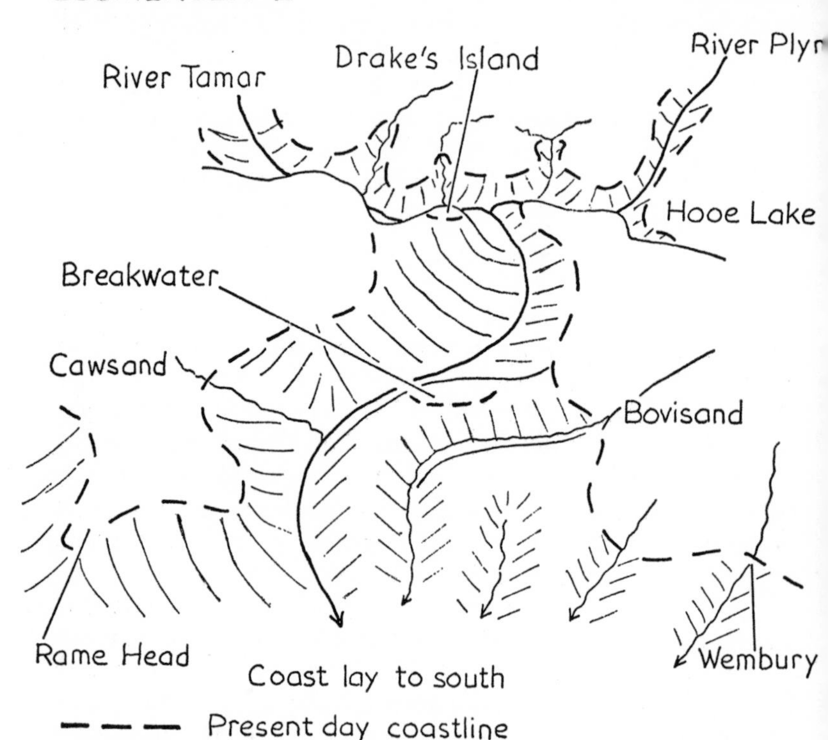

The Tamar occupies a long-standing course between Dartmoor and Bodmin Moor, its history still only partly understood. Like the Plym, it then flowed through a valley some depth below its present estuary. Nearing Barn Pool and Mount Edgecumbe, it, too, turned through the limestone outcrop, deflected by a ridge of high ground barring its way to the sea and crowned by a volcanic knoll—Drake's Island. The Tamar flowed round three sides of this knoll, collecting the waters of the Plym in the process.

The Plym-Tamar junction lay just beyond Mount Batten breakwater and, after it, the Tamar passed between two hills, now the Mallard and Winter shoals of the Admiralty charts, while the valley between is known as the Smeaton Pass. The Tamar ran sw after its great loop round Drake's Island to make another around a ridge extending out from Bovisand—the ridge on which the breakwater stands.

Nowadays the line of this old valley through the Sound is revealed by the courses of ships entering Devonport. For them it is the deep-water channel, skirting the western end of the breakwater and making for the narrows of Smeaton Pass. Safely through the pass, they turn close in by the Hoe shore and move north of Drake's Island to enter the Hamoaze.

It was 10,000 years ago when the sea came back to play a direct role again in the history of Plymouth, in a new increase in level known as the Flandrian transgression. Creeping higher through the Tamar valley, it gradually engulfed the Sound, rising above the level of the hills by the Smeaton Pass and flooding up the present estuaries of the Tamar and Plym. The vast branching waterway it formed is called a 'ria'. In outline, it looks tree-like with the main channels as the trunk and the numerous little creeks the branches. Similar

FIG 2

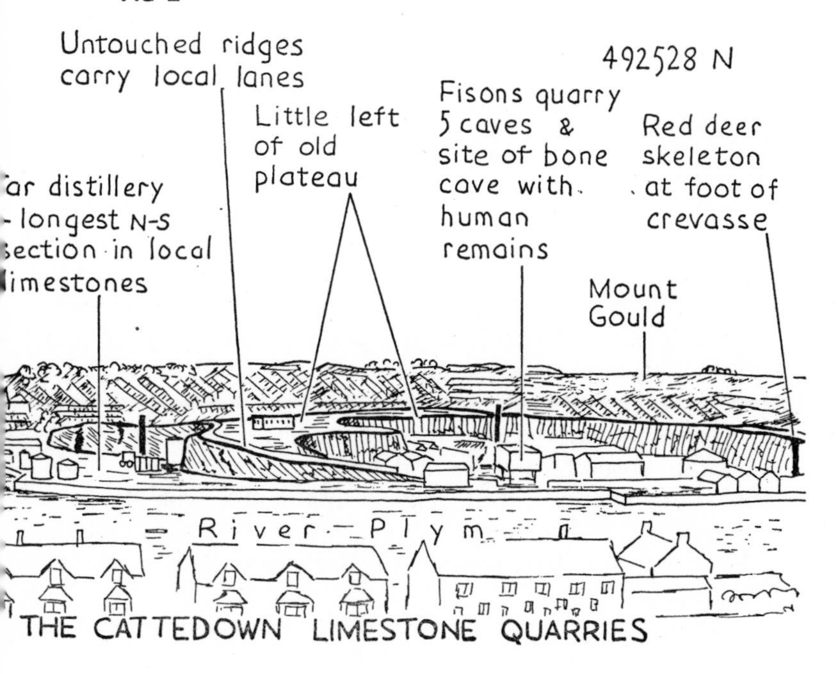

examples exist all along the south coast of Devon and Cornwall in the flooded valleys of the Exe, Teign, Dart and Yealm; at Kingsbridge, Fowey and Falmouth.

Modern but larger examples of the submergence of the Plym can be envisaged. Imagine Cheddar Gorge half-submerged and its surrounding plateau extensively quarried—an exact parallel. The sea, gradually deepening between the limestone walls, created a new estuary which the river Plym has been constantly trying to regain. Dropping vast amounts of silt and sand in the channel, it has buried all the lower walls of the gorge beneath its bed, flooding and sealing off the lower cave systems. The task was helped by the waste gravel brought down from the Dartmoor tin streaming, and by the weaker tidal currents compared to the Tamar where the channel is more easily scoured.

Quarrying has destroyed the original cliffs of the estuary and Cattedown is now filled with industrial sites. The longest north-south section through the limestones in the district is the face of the quarry now occupied by the Tar Distilleries. Students of Plymouth history can probably thank the presence of the Citadel for preserving the eastern end of the Hoe from a similar fate. Only the West Hoe area of that ridge was quarried for building stone and lime-burning.

CREEKS AND HISTORY

The part played by the estuaries in the history of the district can never be fully assessed. Mount Batten was a trading station in Roman times, and in the prehistoric period many migrants must have landed in the estuaries and followed the ridgeway from Mutley Plain to Roborough and Dartmoor. The record seems to grow richer from the middle ages—partly because documentary sources are more numerous and because of increased population.

Up to the late eighteenth century water transport was far more important in daily life than it is today. Roads were poor in the south-west and in Plymouth long diversions around the creeks were necessary, while fords such as Efford (Ebb-ford) and ferries (Oreston and Cremyll) had an early importance.

Before tin-streaming became so active on Dartmoor, ships

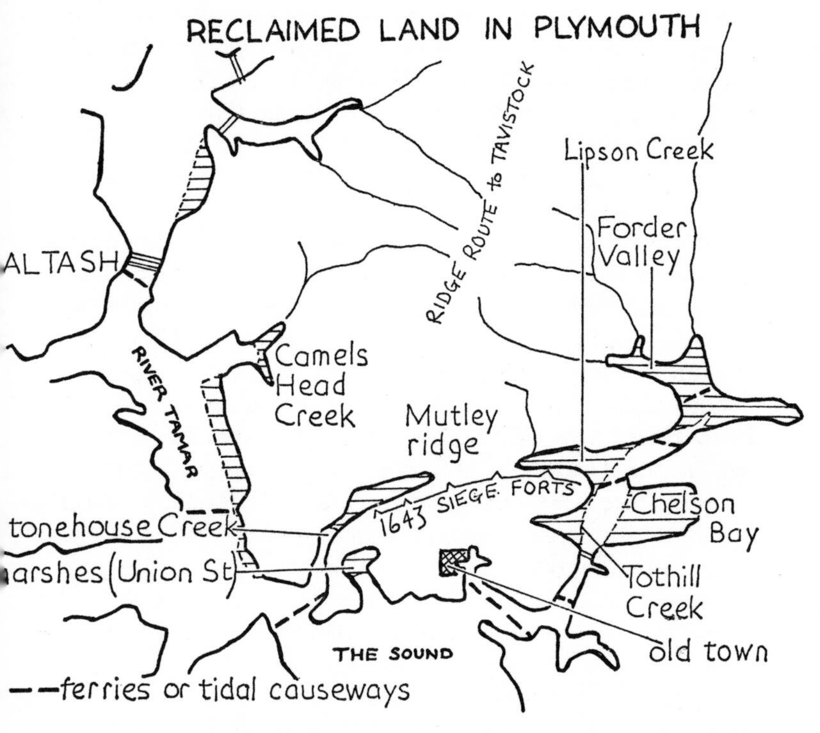

could sail well up the Plym towards Plympton. The tinners' waste gravel soon choked the channel and measures such as the Plymouth Leat (1591) were proposed—partly to help wash the gravel out of Sutton harbour, so far down the estuary that it could have been expected to escape this problem. The leat Act 'for the preservation of Plymouth haven' sounded more effective than it proved, since the leat completed its course at Millbay instead.

The closest relationship of history and landscape ever achieved in the district probably occurred during the siege of 1643-6. To understand why the town was so well-defended, the extent of the creeks at the time must be studied. Two in particular almost encircled the city on the landward side, providing a natural moat—Stonehouse Creek, ending below Plymouth station, and Lipson Creek extending up to Lipson Vale. Between them, the Mutley Plain ridge provided the only 'drawbridge' into the town. A line of forts lay south of ridge and creeks to prevent any successful royalist crossings.

Since then, a good deal of man-made reclamation has taken place, changing the outline of the creeks to their present form —the drainage of Chelson Meadow, the building of the embankment, and the in-filling of Victoria Park, for example. All these efforts have accelerated the process, but natural deposition would have ultimately recovered this land anyway. Map 2 shows the original and present outlines of the estuaries.

Construction of modern bridges to replace the old fords and ferries has also revealed the difference between the old and new river beds. At Laira, beneath the Tamar bridges and the Tavy railway viaduct, the piers pass through 68-87 feet of mud before reaching the rock bed of the old valleys.

One of the paradoxes of the submerged landscape is Barn Pool. Here, just off the Mount Edgecumbe shore and well within the Drake's Island ridge, is the deepest part of the Sound, a site once well up the Tamar Valley. The feature is due to tidal scour removing the estuary muds found elsewhere beneath the Sound.

THE PLYMOUTH LIMESTONES

Plymouth Hoe was one of the coral reefs in the Devonian seas. It grew in a warm ocean rather like the modern Pacific. The reef-forming conditions are described more fully at Lummaton Hill, Torquay, page 51. There were also volcanic eruptions, with material pouring out below water to form pillow lavas. A good exposure of these lavas is to be found in the quarry on the north-west side of Efford Lane. This quarry, at the back of a semi-circle of houses, lies amidst Upper Devonian beds, proving that the eruptions were continuing then. There is a retaining wall built of these lavas along the front of nearby houses (See page 178 for a description of the term 'pillow lavas').

Sometimes the volcanic outbursts reached the surface of the ocean, producing eruptions of ash which rained back into the sea. The ash layers covered the coral reefs, temporarily halting their growth by killing the tiny organisms which secrete the reef material. Soon afterwards the reef life returned, migrating in from unaffected areas of the ocean, and growth recommenced. The ash layers can be found in many parts of the limestones, thin gritty bands of reddish colour. A block taken from Pigeon Cave, Cattedown, revealed at least twenty-seven eruptions over that ocean 350 million years ago. It would be interesting to know the length of time between them but that is impossible. Geologists cannot tell how much the rock has been compressed by later earth movements, so the spacing of the ash-bands has no time value.

The present limestone outcrop begins across the Tamar at Cremyll, where it forms the only Middle Devonian limestone outcrop in Cornwall. It also appears on the north side of Drake's Island but the rest of the island is volcanic ash and

FIG 3

XAMPLES OF FOSSILS FROM THE MIDDLE
EVONIAN LIMESTONES IN SOUTH DEVON

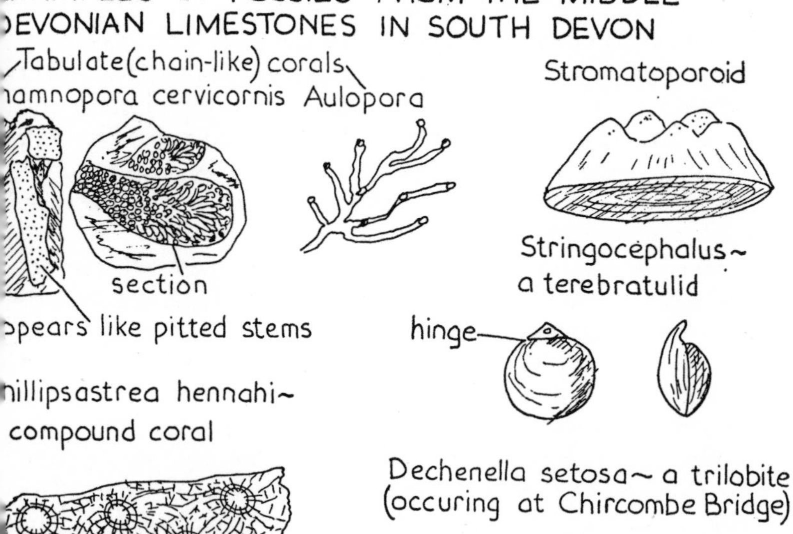

tuff (compacted ash material). These rocks show that Drake's Island stood near one of the Devonian volcanic outlets, although it was not actually a volcano itself.

The limestones pass eastwards from Mount Wise via Stonehouse and the Hoe to Plymstock. They have been extensively quarried around Stonehouse Creek, West Hoe and Cattedown where many weathered faces reveal their fossil corals. Cave formations on the quarry faces show that many old cavities have been destroyed in the quarrying—in the knoll at the south end of Durnford Street for example. The principal modern quarries are east of the river Plym. Beyond Elburton, however, the limestones become patchy until they reach their eastern limit in this part of Devon, almost surrounding Yealmpton village.

PLYMOUTH HOE

Old prints of the Hoe show that little more than a century ago it was a completely natural sea-cliff. Today, the walker must try to visualise it without the complex web of paths and steps. Enough natural cliff is left to occupy his interest, although some of the most interesting features have unfortunately been covered—the 65-foot raised beach is masked by a large retaining wall.

The beach is fully described in records; nineteenth-century geologists went in for amazingly thorough descriptions, although they apparently walked around with their eyes on the ground for they seldom refer to other identifying landmarks nearby and hardly ever record sites on a map. A detailed section is of little value unless it can be pin-pointed on a map.

The 65-foot beach was first revealed when the road around the Hoe was cut in 1816 and its eastern end was exposed in further works during 1884. Its site was described as at the junction of the eastern and western Hoe—probably near the quarry face in West Hoe Park, although other sites could fit this vague reference. Very little granite sand was found in it and only a little limestone from the local cliffs. Most of the material came from Cawsand on the Cornish shore of the Sound—pink felsite, the volcanic rock forming most of Cawsand Beach. Felsite is similar chemically to granite but fine-grained since it cooled rapidly at the surface. There was also some material in the beach from the volcanic beds of Drake's Island.

The way in which beach material could have moved across from Cawsand is well-known and is illustrated in Figure 4. A pebble at A, moved up the beach in the direction of an advancing wave, runs back to the sea directly down the slope of the beach to B. There a second wave coming in repeats the process, and so on. Beach material can move considerable distances in this fashion—Chesil beach in Dorset is built of material from the coasts west of Torquay. So there remains only one unsolved problem about this raised beach—how did the material cross the outflow of the river Tamar?

FIG 4

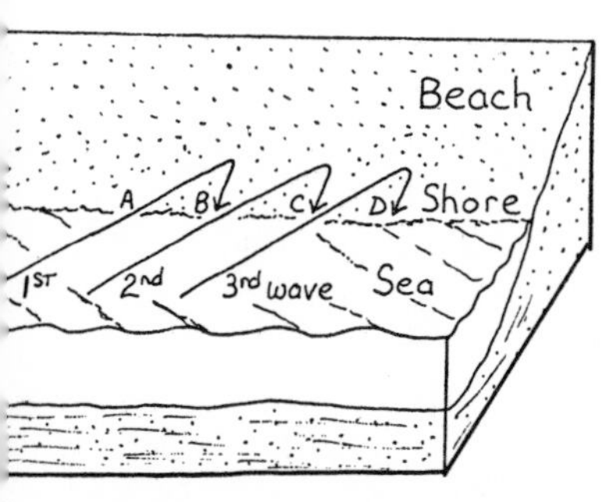

MOVEMENT OF PEBBLES ALONG THE COAST

Pebble moves from A to D with 3 waves. Repeatedly pushed up in the direction they break, it always runs back direct to the waters edge. Great distances can be covered in this way

The 65-foot beach has no counterparts elsewhere in South Devon—probably the remains are hidden behind the later Head deposits and have yet to be discovered.

Fortunately there is another and easily seen raised beach on the Hoe, standing at the 21-foot level. A large shelf-like area of it lies below Madeira Road. It extends about 80 yards eastwards from Lion's Den to a sea-cave near the Royal Corinthian Yacht Club. Viewed from the promenade near the yacht club (Figure 5) where it reaches down to about 14 feet above high water, it seems less convincing than, say, the Hope's Nose beach described at Torquay because it has lost most of its beach material. Fragments of sand and pebbles can be found on it though, and there are two patches cemented to the shelf about 20 yards west of Lion's Den. Here the beach can be traced across the entrance of a large sea-cave to the next area of the promenade. The sea-cave, like the one marking the eastern end of the beach, extends back under

Madeira Road, in this case for a total distance of 240 feet.

Whenever a fault outcrops in the cliffs it is exploited by

FIG 5

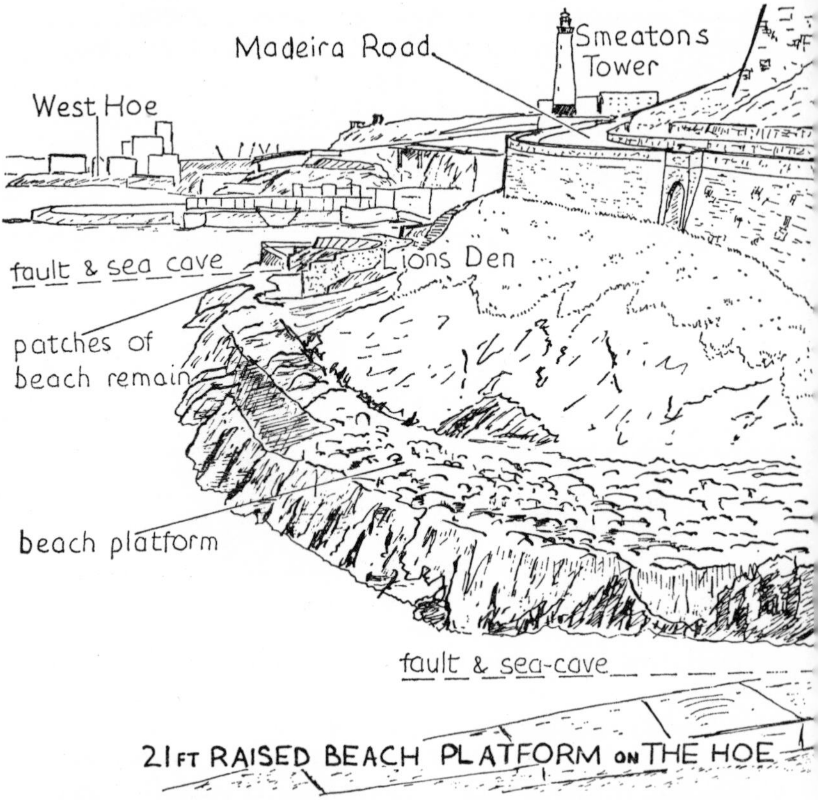

Madeira Road
Smeatons Tower
West Hoe
fault & sea cave
Lions Den
patches of beach remain
beach platform
fault & sea-cave

21FT RAISED BEACH PLATFORM ON THE HOE

the sea. A N-S fault zone accounts for this cave and the sea, attacking along the junction, has been able to widen out a large chamber. Inside, fragments of 21-foot raised beach on the walls date its early existence. The raised beach features were the work of temporary high-sea levels during warm phases in the Ice Age. On the Hoe they were the product of the last of these warm or interglacial phases. The seas receded again with the onset of a new cold phase which locked up the earth's water in growing ice-sheets.

As a contrast to the study of raised beaches, the Hoe slopes have a lot to offer anyone with special interest in fossil corals. Breaking off lumps of limestone is no way to find them, unless you have cutting and polishing machinery, for the fossils are best left to 'weather out'. This is the term for the natural removal of material around them—a slow revelation maybe, but an inexpensive one! The chief corals seen are solitary types, usually exposed in round cross-sections, and branching, chain-like tabulate corals such as *Thamnopora* (Figure 3).

These can also be seen in the polished sections of many Plymouth pavements. Parts of the city are paved with marble! Outside the Central Library and in the square by the Civic Centre abundant fossils can be seen in the alternating rows of white and dark grey slabs. The darker grey stones come from the Pomphlett area of the outcrop.

Because limestone is so easily dissolved and redeposited, many of the fossils found on the Hoe have changed their original condition. Only certain quarries reveal them in their true life positions, usually in the lower thin-bedded limestones like those of Venn quarry, a quarter of a mile north-east of Brixton church. *Alveolites* and *Amphipora* are common there (Figure 25) but permission is needed to visit the site.

West Hoe is low-lying today because of the extensive quarrying and lime-burning which occurred there in the last century. Part of the quarry face can be seen in West Hoe Park, where the massive nature of the Hoe limestones and further cave entrances are revealed. The limestone here is detrital, meaning that since its original formation it has been fragmented, washed away and deposited at a new site. Traced across the road to Pebbleside beach, the beds near the paddling pool reveal sheet-like zones of stromatoporoids.

Along the coast path behind West Hoe Terrace there are more fragments of raised beach just beneath the sea wall and the limestones here are more varied. Thin-bedded pink zones alternate with massive white beds. Solution effects are noticeable in the hollowed surfaces near the shore. At the eastern end of the terrace the massive beds project seawards, while the thin-bedded layers have been eroded away to form sand-filled gullies.

From West Hoe the panorama across the Sound is remarkable. The walls of Tinside bathing-pool provide a convenient sight-line for breaks in the cliffs associated with the 24-foot raised beach on either side. Figure 6 shows the relationship between surface levels and rock type visible on the east shore.

The Meadfoot Beds and Staddon Grits dominate the view, revealing the edge of a great coastal plateau which runs inland from the wireless masts and the great wall of Fort Watchhouse towards Down Thomas and Staddiscombe. In the centre of the view a middle level is seen in the less resistant slates around Jennycliff and Fort Stamford, and finally, in the volcanic tuffs near the officers' mess at Mount Batten, very low cliffs indeed.

FIG 6

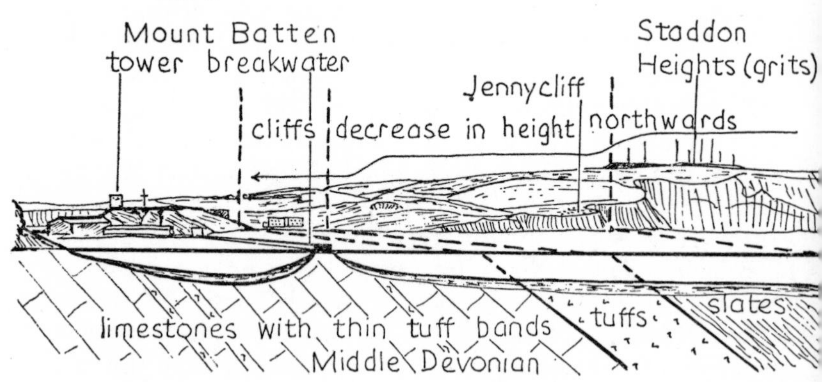

EASTERN SHORE OF PLYMOUTH SOUND
CLIFF FORMS & ROCK TYPES

MOUNT BATTEN

Crowned by its little round seventeenth-century tower, Mount Batten also has a marked erosion level on its summit. This surface can be traced for some distance through the limestone outcrops of the estuary—around Clovelly Bay, the entrances to Hooe Lake, Cattedown, and the ridge followed by Plymstock Road.

Seen from below the Citadel, Mount Batten illustrates bedding, faults and the difference between true and apparent dip. Its limestones seem to dip westwards towards the Mount Batten breakwater but this is only because the rock face in which they are seen runs in that direction. A three-dimensional view would confirm that they really dip more towards the back of the hill, to the SSW.

FIG 7

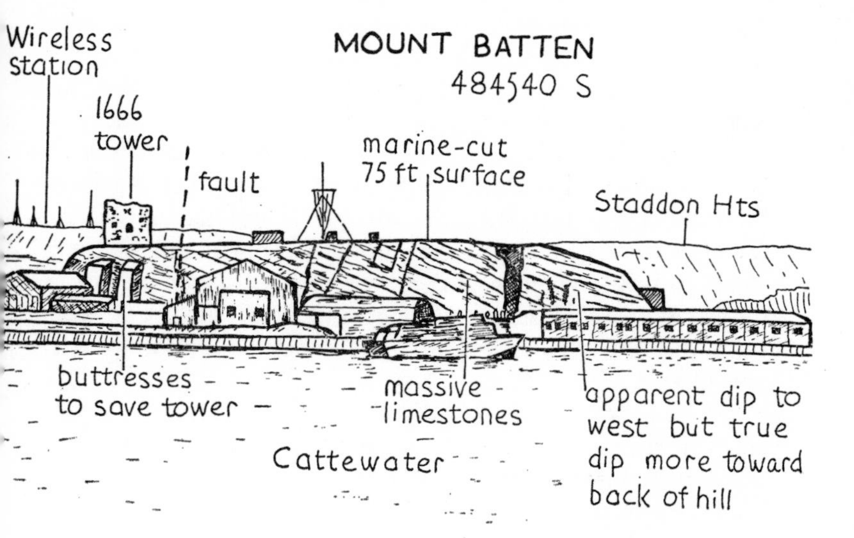

One of the faults in the hill, passing close behind the old tower, caused a considerable problem. The tower had a water-tank inside and the weight, aided by the fault behind and the too close quarrying below, threatened to bring that part of the hill down. Massive concrete buttresses now preserve the tower, not only as a monument of history but also to the value of a little preliminary engineering geology!

When the lowest raised beach on the Hoe was formed Mount Batten was an island. The low-lying tuffs to the south were covered by shallow water, a close parallel of the modern Drake's Island-Mount Edgecumbe relationship.

PLYMSTOCK AND WEMBURY

Before the explorer crosses Laira Bridge to Plymstock he should visit the Cattewater Road exposures, between the bridge and Plymouth power station. The northern boundary of the limestones is exposed at the bend in the road. Beneath it, on the inside margin of the bend, thin early developments of limestone can be seen in southward dipping and highly cleaved slates. The part of the road leading to the power station is entirely in limestones, the higher beds dark grey in colour with crinoid, coral and broken stromatoporoid remains. The junction of limestone and slate can be traced to the outer side of the bend where there is a good view of the Plym estuary. Beyond the reclaimed Chelson Meadow and Embankment areas, ridges of grits and shales at Efford and Saltram enclose a distant view of the granite at Lee Moor.

Alternating beds of limestone and slates can be seen again across Laira Bridge along the road into the Portland Cement Manufacturers' quarry. This quarry is the largest modern working in the limestone belt along the north side of Plymstock and over 500,000 tons a year of high-grade cement are made there in the complex of crushers and silos. The works supplies Cornwall, Devon and the western parts of Somerset and Dorset.

On the south side of Plymstock there are outcrops of volcanic ashes or tuffs, part of a long exposure extending into the South Hams. They can be seen in the dip where the Hooe road crosses Radford Valley. Broken samples of these red rocks reveal crystalline faces. They make rather difficult soils, heavy in winter and brick-like in summer.

The Radford Valley flows into Hooe Lake but it is best to go on into Hooe itself to see this feature. The lake, really a creek of the Plym, is enclosed behind arms of much quarried limestones. A N-S fault provided it with an exit through them to the Plym. The inner shore of the lake is much gentler, being cut in slates, but since there are no exposures here it is best to go up to Jennycliff to see these beds.

Thin, silty bands with ripple bedding can be seen 100 yards north of the beach path. In the base of Jennycliff low-angle faults can be traced, probably northward thrusts, and combined with a second series of nearly vertical faults they have produced block-faulted structures.

Midway between the beach path and the track down from the caravan site the cliffs falsely seem to contain a raised beach deposit. Minerals washing out have been colonised by mosses to form green stalactite-like forms. The geological deception is completed by the storm-thrown pebbles embedded in these spongy masses.

Continuing up the road to Staddon, the walker can renew his study of the Sound and at Bovisand notice the still youthful and steep forms of the few valleys in the high grit

FIG 8

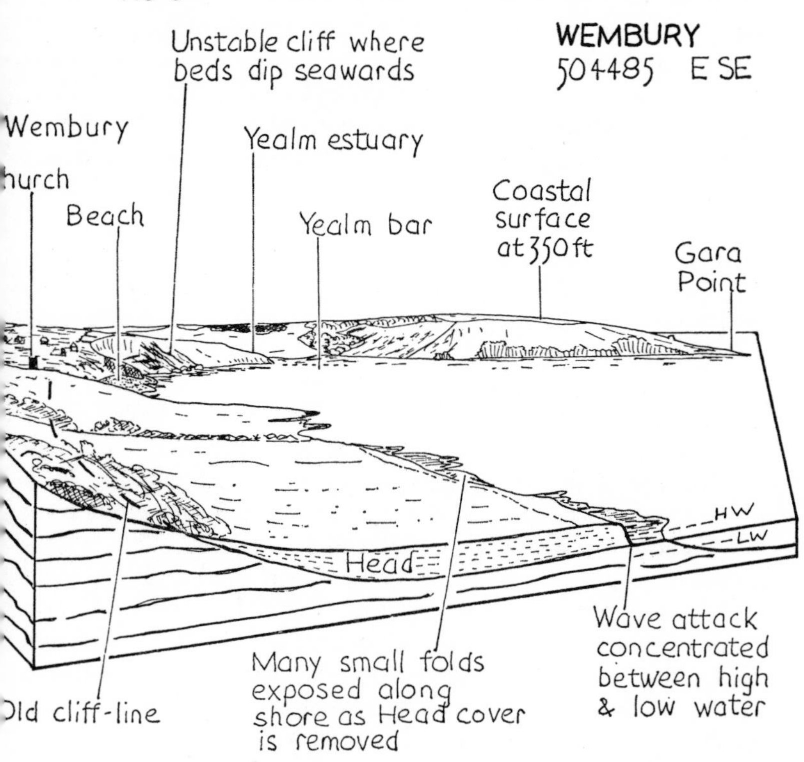

plateau. The streams obviously find the grits hard material.

At Bovisand beach and along the coast to Heybrook Bay and Wembury the typical Head deposits of the South Hams begin (Chapter 2). Earthy cliffs appear, with solid rock in platforms below them and in higher crags inland. Near the entrance to HMS *Cambridge* at Wembury Point the view towards the Yealm estuary reveals a remarkable extent of Head. A gently sloping field, extending nearly as far as the barely visible tower of Wembury church, marks the upper surface of the Head. Its inland hedge-bank marks the junction with the solid Dartmouth Slate outcrop.

YEALMPTON

The Yealm estuary is another of the many South Devon estuaries formed by the submergence of an existing valley.

Near the innermost reaches the Plymouth limestones outcrop again. Almost surrounding Yealmpton village, they exhibit some remarkable colour changes. In Kitley Park, on the west side of the village, they used to be quarried as Kitley marble, a green-tinged rock due to the mineral glauconite. Glauconite forms today in marine areas near the edge of the continental shelf, the shallow water area fringing the continental landmasses.

At the opposite end of the village, in Eastern Torrs quarry, there is a very pink limestone. Easily distinguished in modern walls and garden paving of the Plymouth area, the quarry from which it came was the site of an important bone cave, discovered in 1954. Deposits had accumulated below a hole in the roof, and animals fell into the system in the same fashion as at Joint Mitnor, Buckfastleigh (page 87), although at Yealmpton they were then washed on some distance by a stream. The deposit was of Last Interglacial age, containing hippopotamus, cave lion and hyaena bones.

There are another sixteen caves south of the village, and to reach them the walker must make for the river and immediately after crossing the bridge turn sharply westwards along the lane which follows the Yealm towards Puslinch. At the footbridge over the river are several limekilns, standing by the entrance to the quarry containing the caves.

Near the quarry mouth are the once-gated entrances to the old Kitley show cave. All round the quarry partly destroyed cave formations and other entrances can be seen. The adventurous can walk through the show cave starting from the entrance beneath the steeply inclined exposure and turning left just inside, eventually coming out at the higher door. Unfortunately, the best features in it have been wantonly destroyed.

East of Yealmpton, the limestones die out and the volcanic tuff outcrops become more important. In the ridge south of the Ermington-Avonwick road they are known locally as dunstones and make valuable farmland. Limestones and tuffs appear in the cutting on the Modbury road east of Sequers Bridge but it is an unsafe place to collect specimens! It is worth turning off at the far end of the cutting and taking the lane to Oldaport and Kingston, for as the one-inch map reveals, near the village felsite and dolerite intrusions break up the pattern of Lower Devonian outcrops. Around Great Torr these intrusions form amazing rock piles, jutting up in the midst of the smoothly rolling landscape.

Climbing the hill south of Oldaport, park by the first junction on the left and walk back about 75 yards to the outcrop of Torr Rock. Almost vertically-walled and marking the field boundaries, its mass is formed of pinkish felsite and much white quartz.

For the explorer who wishes to go on into the South Hams the district offers few other inland exposures and it is best to make for the coast. Some outlying patches of New Red Sandstones can be seen near Thurlestone, but a detailed description of South Hams features is given for the east coast in Chapter 2.

Chapter 2

Slapton and Start Point

The South Hams, lying between the estuaries of the Dart and the Plym, is famous even in South Devon for its rural beauty and agricultural value. Although it is never more than 700 feet above sea level it is, as old guidebooks quaintly say, full of hills and valleys.

Geologically it resembles a sandwich with a filling of less resistant rocks between two harder and more elevated outcrops. The South Hams district reaches its greatest E-W extent

MAP 3

(Generalised from maps of the Institute of Geological Sciences with permission of the Director)

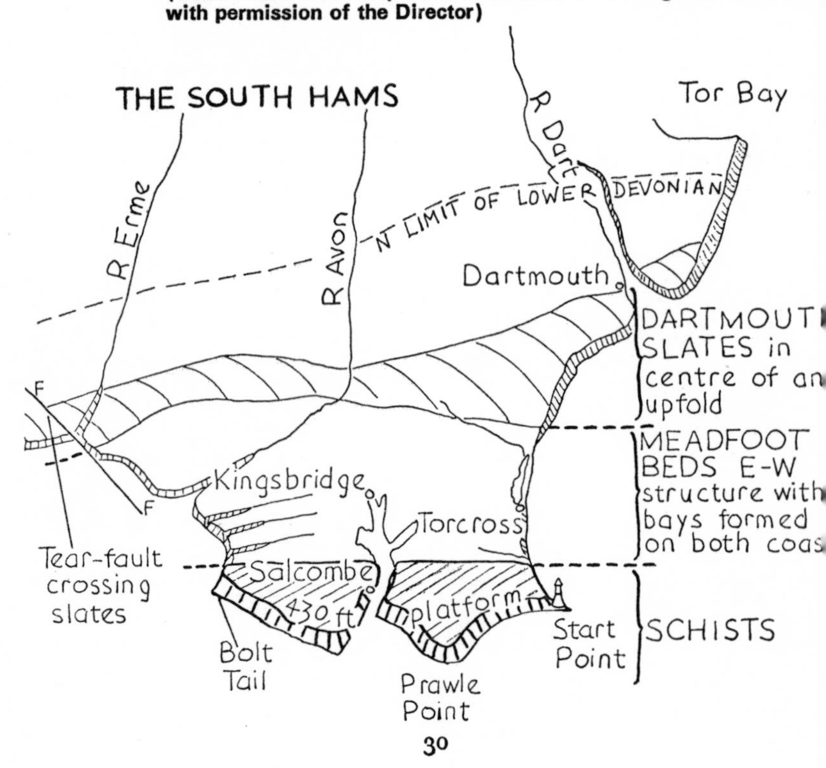

in the northern Dartmouth Slates series. Southwards, the Meadfoot Beds represent the filling and the area narrows here where the broad bays of Bigbury on the west and Start on the east have been cut out in these rocks. A more elevated area of metamorphic rocks completes the sandwich on the southern side, and causes the South Hams to broaden out again between Bolt Tail and Start Point.

One of the recurring problems of Devonshire geology is the lack of inland rock exposures, but in the South Hams this can be overcome by walking along the fine N-S coast of Start Bay. This coast cuts a right-angle section across the rock sandwich. Sweeping southwards in a long, graceful curve, it possesses one of the longest shingle beaches in Devon. Continuous at low water and broken only by Dun Point and Tinsey Head at high water, it extends nearly six miles from near Strete through Torcross and Beesands to end south of Hallsands village.

The formation of this beach has made a striking alteration to the scenery of the coast, sweeping across the old irregular outline and sealing off former bays. These have now become the freshwater leys of Slapton and Widdicombe, near Beesands. They form protective barriers preventing the sea from cutting cliffs along those parts of the shore.

The best way to see this coast is to make a full day walk along it from Slapton to Lannacombe, but it can also be divided into convenient short sections by car journeys between the main villages. Parking is also possible near Start Point and at Lannacombe beach.

THE DARTMOUTH SLATES

Folded with all the Devonian and Carboniferous rocks of Devon about 270,000,000 years ago, the Dartmouth Slates now form the centre of an up-fold, with the Meadfoot Beds marking its southern limb. The Dartmouth Slates are the earliest Devonian rocks visible today, but the exact nature of the beginning of Devonian times cannot be proved because their base has never been found.

North of Blackpool Sands, the cliffs are flat-faced where the beds of slate have been folded into a vertical position. The numerous blunt-nosed little headlands and very small coves show how difficult it is for the sea to erode this coast. Towards the beach the cliffs become very earthy, probably from material washing down the valley side. Protection has been necessary here and blocks of hard greenish dolerite from the South Hams quarry at East Allington, and some pink limestone blocks from Plymstock, Plymouth, were tipped at the back of the beach in 1966.

The small promenade is an interesting feature because it marks the eastern boundary of a submerged forest. Beneath the sands lie layers of clay and tree stumps which have been exposed on several occasions when the sea has temporarily swept away the covering material. Easterly gales pile the sand up at the western end of the beach and westerly ones vice versa, and any regular visitor to Blackpool Sands will know how much its beach varies in shape. In February 1869 the gales removed the western sand revealing a clay deposit with tree stumps and other remains. The clay was generally brown in colour, with blue layers nearer the shore. It was revealed again in April 1881 when the central part was laid bare. With other submerged forest sites in Tor Bay and at Bovisand, near Plymouth, it is evidence of recent minor changes in the level of land and sea.

Inland from Blackpool Sands there are some nearly impossible gradients along the lanes. The Dartmouth Slates are a resistant rock group and the streams have only managed to cut narrow valleys in them. The ridges soon reach heights of 450 and 550 feet, both being old erosion levels cut perhaps by late Ice Age seas (see page 37). Even the main coast road to Kingsbridge has to climb steeply southwards to Strete.

The roadside walls and most of the houses in Strete are built of the local slates and shales, but more recently repairs have been made with limestone which has to be carried nearly thirty miles by road from Plymouth. Here history repeats itself, for the same limestone used to be sent by sea to the South Hams coast for lime burning, a trade at its peak in the early nineteenth century. The modern use of it shows how good stone can still find distant markets today. It is also evidence that the local quarries have not worked for some years now. Even when they were in use these quarries never yielded slate good enough for roofing purposes, and the fourteenth-century tower of the Collegiate Chantry at Slapton had to seek stone for its walls from slate quarries farther west at Charleton.

South of Strete the main road provides an ideal vantage point to admire the great sweeps of shingle beach and to compare its regularity with the original outline of the shore—now forming the inland margin of Slapton Ley.

EARLY DEVONIAN LANDS AND SEAS

Quite naturally, geologists have paid a great deal of attention to the Dartmouth Slates, trying to find out what

FIG 9

YCLOTHEMS – developed by a meandering river

attern of sediment at each bend

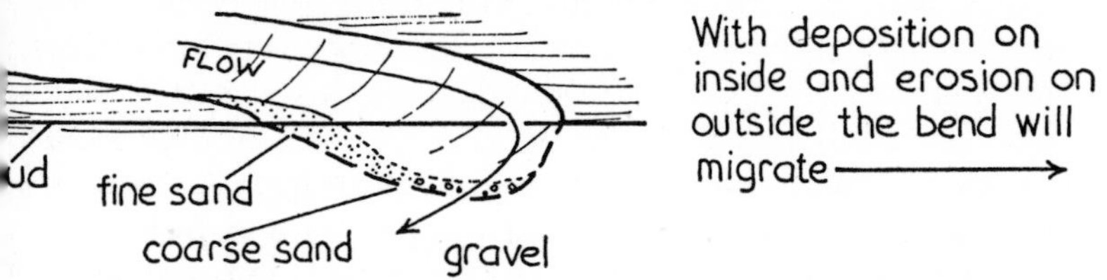

As each bend meanders to and fro it creates continuous beds of gravel, sand & mud

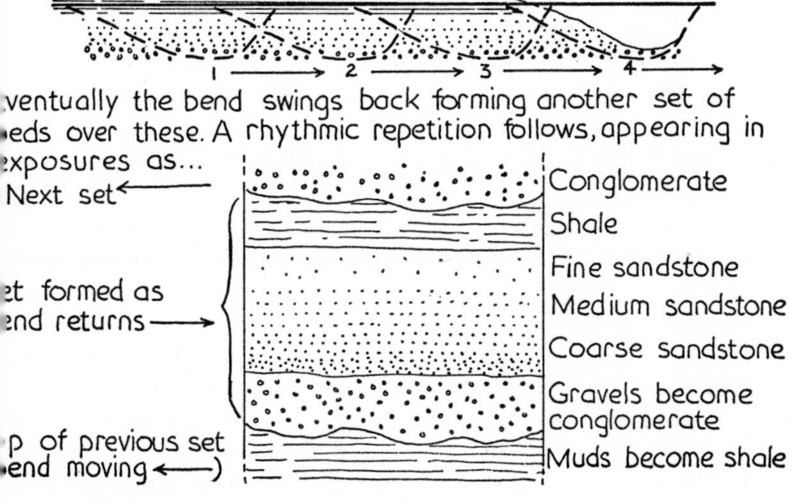

conditions were like when these earliest Devonian beds were formed.

In Devonian times the sea was gradually encroaching northwards over England. Sometimes there were pauses in its advance and even periods of short retreats, so in some localities alternating land and marine deposits accumulated. For most of the Devonian period, however, the coast lay over the modern Bristol Channel area, and north of this the Old Red Sandstone rocks were forming in arid land conditions. The rocks of North Devon are typical examples of marine deposits formed close to a coast, while those of South Devon were mainly formed in deeper waters.

Dr D. L. Dineley has solved one outstanding question about these South Devon rocks—were there any rocks here which were not marine and which might have formed before the submergence of this area occurred? Obviously the Dartmouth Slates had to be studied in detail, being the earliest beds. After making the journey from Blackpool Sands to Strete Gate, the student will appreciate how difficult this was. The slates also included coarse conglomerates, coarse and fine sands, shales and many large quartz veins. Added to this, the severe folding of this area complicated the picture and meant that one layer might be repeated many times in a complex section.

Detailed studies led to the interpretation that the Dartmouth Slates were formed during a land phase of low-lying, flood-plain conditions. They accumulated in an area which was almost at sea level, inter-tidal in parts and crossed by numerous braided river channels. These channels changed position, wandering sideways and building up the level of the lowland exactly as rivers and their deltas do today. The rhythmic sequence of deposits produced by this is termed a cyclothem.

Each cyclothem follows a sequence of coarse conglomerate, at the base, and then coarse sand, medium sand, fine sand and mud (Figure 9). The coarse conglomerate represents the material in the bed of the river, the sands being formed in the quieter outer areas of the channel and the muds on the flood plain beyond. As the channel shifts off sideways and raises its beds by its own deposition, so its conglomerate sand and mud layers begin to form diagonal bands. Hence the varied nature of the Dartmouth Slates—their name really is a misnomer! Since each layer cuts across time in its gradual extension, the beds which are formed are called diachronous deposits.

Thus it was after the Dartmouth Slates phase that the coast made its move north towards the Bristol Channel and South Devon was submerged to continue the rest of its deposition in a marine environment. The Meadfoot Beds represent the first fully marine deposits in the Lower Devonian of the South Hams.

SLAPTON SANDS

The shingle beaches of Start Bay include a variety of rock types. There are pieces of local material, grey slates and white quartzites, fragments of schists from the metamorphic area with diffused mica shining in them, and many shell remains. These local pieces are all white or dark grey and green in colour, whereas the beach is strikingly golden in appearance. This is because it contains so much flint. Flint nodules are found in the chalk beds of the Cretaceous period but there are no rocks of this age in the South Hams. In fact the nearest land exposure of this material is at Beer in East Devon (Chapter 10), thirty-five miles away by sea and in the opposite direction to the eastward movement of material in the Channel. The flint on the South Hams beaches is dredged up by the sea from Cretaceous rocks in the floor of the English Channel. Geologists have found that the chalk extends westwards beneath the sea towards the Eddystone Reef, south of Plymouth. The chalk rocks may have occurred this far west on land, too, at one time thinly covering Dartmoor. Their effect on the moor will be described in a later volume.

A second 'foreign' material in the beach comes from Dartmoor. Fragments of granite can be found, brought into the northern part of the bay by the river Dart.

Slapton Sands is a most soundly constructed beach, partly because of slight changes in land and sea level which have increased its height. It probably grew southwards as a spit across the mouth of the river Gara, a growth which followed the tidal movement in the bay (see Map 4). Once it had completed the barrier down to Torcross, the ley enclosed behind it became freshwater and is now drained by a tunnel

FIG 10

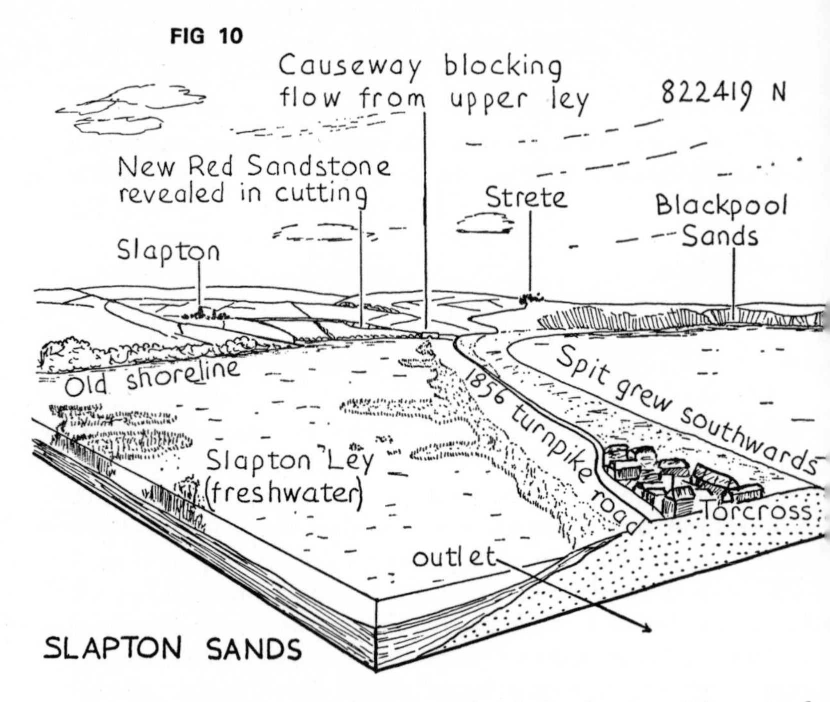

through the slate headland south of the houses. The raised beach of Slapton Sands now stands 20 feet above sea level and gained additional strength from the turnpike road built along it in 1856. The turnpike era was notable for new routes with easy gradients. If the modern traveller avoids the beach road and tries to journey from Stokenham to Slapton and Strete by the inland lanes he will soon picture the old difficulties of travel and isolation.

The causeway leading from the beach road into Slapton was built as part of the same improvement. Unfortunately, it has slowed down the flow of the river Gara southwards through the ley and caused it to drop its load of sediment in the upper part which is now nearly silted up.

The western shore of the ley is in the Meadfoot Beds, the first marine deposits in the Lower Devonian, and by the left-hand gate west of the causeway a small fold in buff-coloured shales can be seen. The Meadfoot Beds are mainly slates, their usual colours red, buff or grey. They are less resistant to erosion than the Dartmouth Slates and rise inland to form ridges at about 320 feet, a level which is also attributed to a former but much more recent sea floor.

THE PLEISTOCENE SEAS

During the later part of the Ice Age, sea level was much higher than it is today, and a good deal of Devon must have been under water. It is debatable how far the sea rose, but it was certainly as high as 430 feet above its present level and may have been as much as 690 feet. In that case Dartmoor would have been an island. Theories about this will be discussed in the author's *Geology Explained in Dartmoor and the Tamar Valley,* to follow this volume. Whatever its maximum height the sea certainly created an extensive erosion surface at 430 feet planing across much of Cornwall, the Start Point-Bolt Tail metamorphic rocks and other areas of Devon. All along this coast and from many other vantage points in the South Hams, this 430-foot level is seen as a striking elevated plateau. A stranger to the district can readily identify it by the Start radio masts standing on its surface.

This is another reminder that the South Hams is a geological sandwich—the resistant northern and southern areas reaching 430 feet or more, the intervening weaker Meadfoot Beds lowered to 325 feet. Other levels occur at 400, 375, 300 and 280 feet. Fragments of them can still be traced along the hillsides and they mark pauses as the sea returned to a lower position, cutting a 'staircase' of levels on the South Hams landscape.

The sea continued its retreat to a level below and beyond the present coast. The floor of Start Bay is now 36 to 50 feet below sea level, but it ends in a high submarine cliff at its outer margin. This probably marks the old coastline and now lies 150 feet below water level. It was during the severe cold phases of the Ice Age that a great deal of the earth's water was locked up in the growing ice sheets and sea levels were lowered. Corresponding rises of level occurred in the intervening warm phases when the ice caps melted and retreated. A rise of this type brought the sea back up from its low level in Start Bay to a height 150 feet above modern level, where it was able to bevel off the New Red Sandstone rock outlier near Slapton.

From late Ice Age (Pleistocene) times lower benches at 65, 24 and 14 feet mark the most recent movements of sea level. These can be traced along many parts of the Devon coast (see Chapters 1 and 3). In the South Hams there are some good examples in the metamorphic area, south and west from Hallsands.

SLAPTON

The Anglo-Saxons were good practical geologists and they chose a village site with ample water supply. Names such as Church Wells Cottages recall their old resources from wells and shallow springs. An outlying patch of the New Red Sandstones, more familiar in the red cliffs of Teignmouth and Budleigh Salterton, forms a natural aquifer around the village of Slapton.

FIG 11

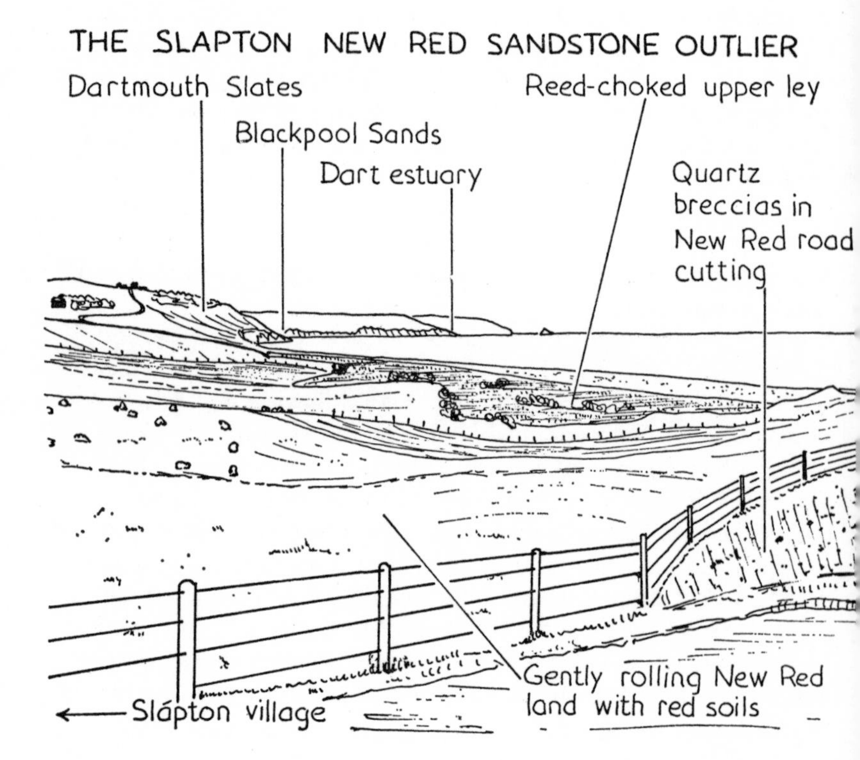

The sandstones create smooth rounded slopes and valleys and the scenery near the village is typical of parts of the Teign and Exe valleys. Pebbles are found in the sandstones—in the road cutting above the causeway there are pieces of white quartz. Broken from the older Devonian rocks of the neighbourhood, they were washed down with the sands to become part of the newer rock.

The Danes in their raids on Devon are reputed to have landed in Ireland Bay, sailing over the position of Slapton Sands as they are known today. If this was so, it means that the sands have accumulated since the tenth century. The name of Darnacombe Valley south-west of the village commemorates the raiders and no doubt they attacked Slapton and possibly Slapton Castle, too. Slapton Castle, an oval earthwork, straddles the lane from Stokenham. Most of the ditching around the six-acre defensive site has been destroyed by ploughing but large flint pebbles can be found. Coming from the sands, these may have served as sling stones at some period in the castle's history.

TORCROSS TO HALLSANDS

Torcross stands partly on Slapton Sands and partly on one of the dark slate headlands around Dun Point. Northwards from the village, the sands have been given protection from cars by rows of small posts. Students from the Slapton Field Studies centre regularly study the re-establishment of salt-loving plants here. Sea spurge and rest harrow are prominent varieties. South of the village the cliffs are high, folded and cleaved. There are much thicker beds of slate, dark clayey layers and some once molten rocks. The latter can be picked out in the cliffs as greenish bands containing soft green chlorite and flakes of glistening white mica. The molten layers were probably dolerite but, like the surrounding slates, they have been altered and rotted, brown bands of oxidised iron minerals making recognition difficult. Since both rock types have experienced the same process they are probably contemporary.

Standing back on the beach the student can see that the rotted brown bands are often parallel to the massive veins of white quartz running through the cliffs. The rotted bands

FIG 12

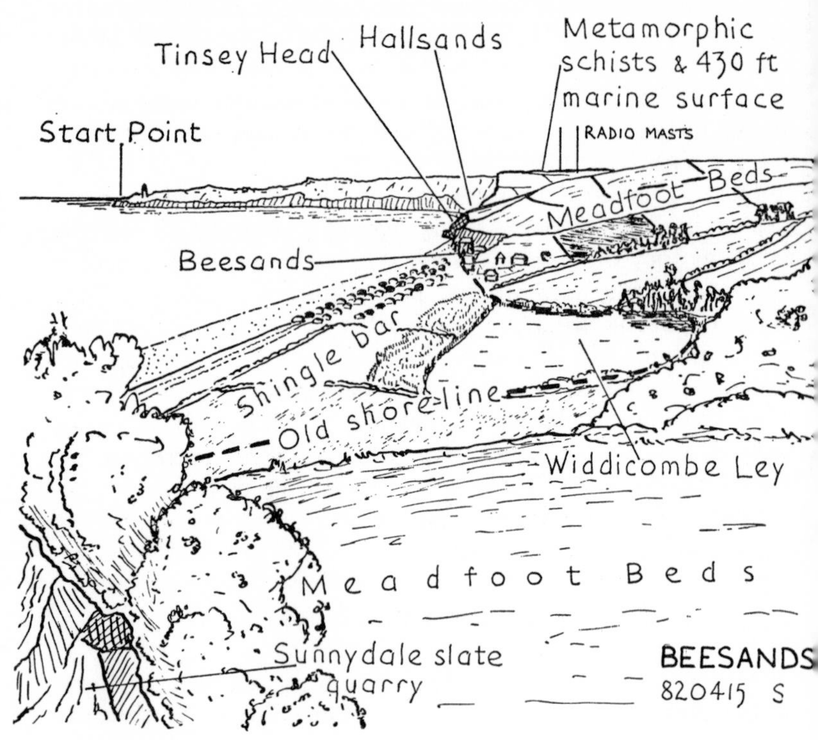

are the only clues available about the original bedding in these highly folded cliffs.

Like the dolerite, the massive quartz must be contemporary with the slates, but there is a second type of quartz veining here, too. These are more recent, thin, irregular veins which run haphazardly across the beds and represent later filling of small cavities and cracks produced by stress in the area.

Climbing the steep paths from Torcross, the whole length of Slapton Ley can be admired to the north. By the gate into the first field there are some large shale slabs of a type produced in quarries north of Stokenham. They can still be seen in field boundaries around that village and at Torcross. The track skirts the hidden edge of the large Sunnydale quarry before descending again to Beesands.

Sunnydale quarry is almost completely hidden in Dun Point headland. The concealed entrance can be found a few yards north of Beesands beach, and was just wide enough for a cart to enter. The quarry was worked for slate and its variation meant that it was useful for roofing as well as building. Since it closed, much of the old cart-track leading to the beach has been eroded away by the sea.

All along the Torcross-Hallsands coast the scenery alternates between ridges with cliffed ends and valleys with enclosed leys. Their east-west direction reflects the structural grain of the South Hams. Streams flowing down these coastal valleys have gradually eroded back into the landscape at the expense of the streams draining towards the Kingsbridge estuary to the west.

At Tinsey Head, the internal structure of the ridge between Beesands and Hallsands can be seen in detail. The beds have been folded into a vertical position, although the slates are not as highly cleaved as those near Torcross. Some crinoid stems can be found in thin bands of dark limestone just north of the head, and white quartz veins are again a prominent feature.

THE METAMORPHIC BOUNDARY

The Meadfoot beds at Tinsey Head are the last exposure of the Lower Devonian. At Greenstraight, just north of Hallsands, there is an abrupt junction with metamorphic schists. It is not a spectacular feature to look at, the cliff at this point being a low one. Scenically, the best feature along it is the course of the Bickerton Valley, following it down to the shore.

This boundary is still something of a puzzle. It has its counterpart in Cornwall, where there is a similar faulting on the margins of the complex Lizard area. In the Lizard, there is also a complicated crush zone north of the fault, known as the Meneage crush zone. The South Hams metamorphic area is more simple but the two areas have been tentatively linked and the features interpreted as part of a great thrust-fault. Slicing through South Devon and the Lizard, this fracture was caused by a thrust from the south.

HALLSANDS AND START POINT

The now ruined village of Hallsands was originally built on

a bench cut by wave action when the sea stood 14 feet higher than today. Where the rock shelf was absent walls were thrown across from point to point and shingle used to fill the spaces behind them. There were thirty-seven houses here before disaster overtook the village in 1917. More recently, neighbouring Beesands was similarly threatened and there in 1962 extra defensive measures were taken, including the dumping of dolerite blocks as at Blackpool Sands. This reprieved a fishing community of at least six centuries standing, one which now concentrates on crabs and where in March the visitor can see the crab-pots being made on the beach from saplings cut locally.

The Hallsands disaster was due to extensive shingle dredging begun in 1897, and an estimated 650,000 tons were taken away for use in extensions to the naval dockyard at Keyham in Plymouth. This caused a drop in the beach level of about

FIG 13

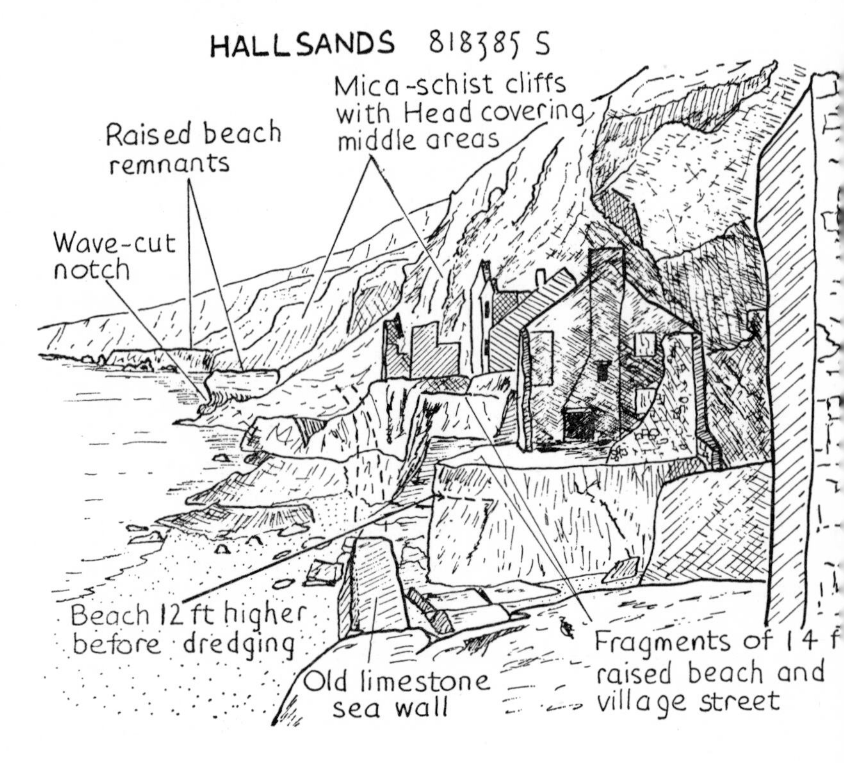

12 feet, leaving the houses without the protective barrier which had absorbed the fury of easterly gales. As a result, the houses on the old mica-schist bench were destroyed in 1917 (Figure 13), despite earlier warnings in 1903 when most of the foreshore disappeared in a single storm. In 1904, the side of the London Inn collapsed one afternoon at tea-time. The people responsible for the dredging did not understand the movement of shingle in Start Bay.

MAP 4

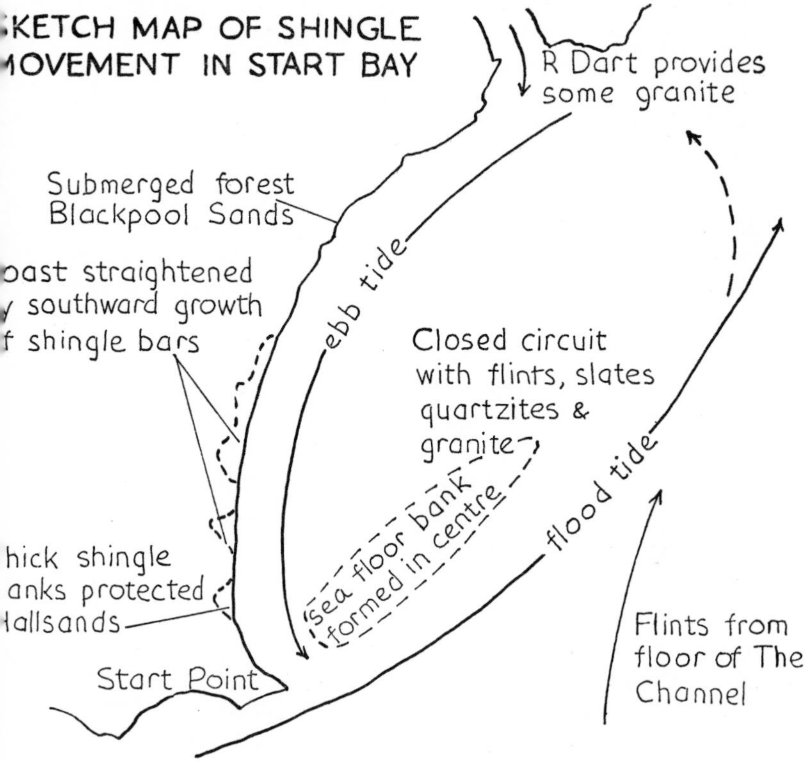

KETCH MAP OF SHINGLE OVEMENT IN START BAY

Whatever the conditions in which the great length of shingle beach accumulated, today the movement of shingle in the area is self-contained. Little new material enters the closed-circuit system. Map 4 shows how the flood tide moving up-Channel carries material around the outer margins of the bay, leaving the ebb tide to migrate it south-west along the beaches, against which the long easterly fetch acts as a bulldozer to pile up the material. The Hallsands dredgers removed a significant part of the bay's shingle reserves, falsely assuming it would be replaced from other sources and thinking that the same conditions which constructed the beaches still applied. Disaster followed. New houses were built for the villagers in the Bickerton Valley north of the old limekiln and the hotel.

South of Hallsands all the high ground is formed of metamorphic schists, rocks with a foliated appearance. There are three types here : hornblende schist (dark green), quartz schist (white streaky), and mica schists (glistening). In the early geological maps they were placed in the Pre-Cambrian period, at least 200 million years older than the Devonian rocks, but geologists now know that they were metamorphosed in Devonian times.

What is certain about the metamorphic area is its effect on the scenery of the South Hams. The elevated marine-cut plateau is visible all over the district. Local streams, following shallow hollows inland, plunge through deep combes near the coast. Along the coasts the schists appear as jagged ridges in Start Point and Peartree Point, but they are not solely responsible for the cliff scenery—other processes have altered its appearance a good deal.

THE GLACIAL HEAD DEPOSITS

The upper areas of the cliffs between Start Point and Lannacombe are jagged solid rock. This is due to the shattering effect of frost action in the intensely cold conditions of the Ice Age. Blocks of rock were loosened and fell down the slopes, and in the brief summer seasons the completely frozen ground might thaw out to a depth of a foot or two. The top layers were then so waterlogged that they slumped and flowed away downhill, carrying away all the stones that had fallen on them from higher up. A lot of earth must also have slipped off the plateau above in this way and poured down the cliff faces. The effect of this seasonal 'freeze-thaw' action was also important on Dartmoor and in the caves at Buckfastleigh. Here on the coast, it altered the original appearance of the cliffs, so that instead of a rocky cliff face there is now a sloping shelf about half way down. This is the upper surface of the slumped earth and stone material. Burying the lower part of the original cliff face, it is known to geologists as 'Head'.

Sea cliffs have now been cut in the Head itself, and here the brown earth and stone layers are in marked contrast with normal rock cliffs. As the stones were carried down the slopes, they descended with their long axis pointing down-gradient so that in these earth cliffs they now jut out end-on. This is a quick means of identifying the Head material.

FIG 14

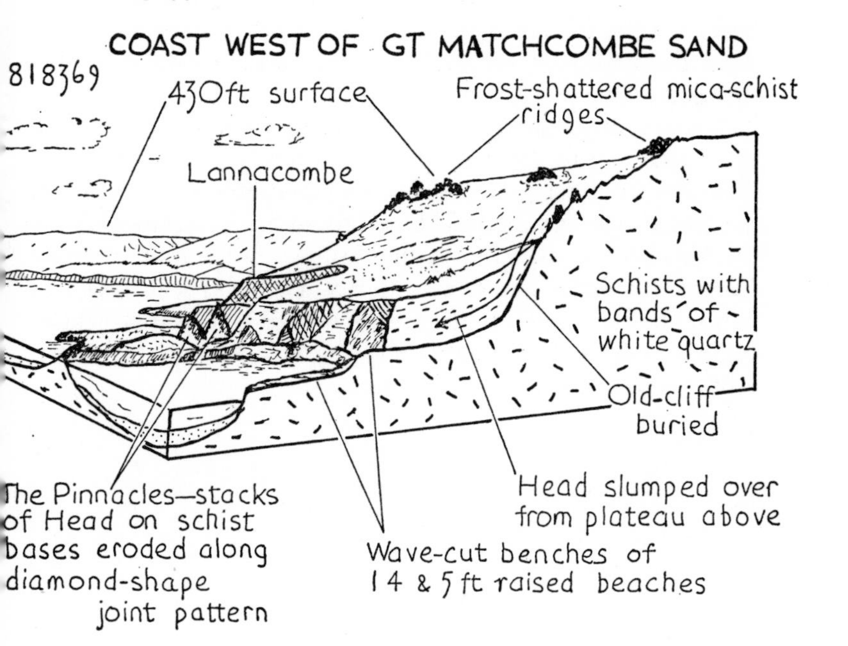

PEARTREE POINT TO LANNACOMBE

Peartree Point, very roughly sculptured by frost action, almost forms an arete on its upper slopes. Across Great Matchcombe Sand, and westwards to Lannacombe and Prawle Point, the great mass of Head forms cliffs reaching 50 feet or so in height. Narrow fields occupy the surface slopes of the Head, with solid rock appearing again above and leading up to the edge of the 430-foot marine plateau.

On the rocky shore below the Head there are two wave-cut platform levels at 24 feet and 14 feet, representing late changes in sea level. Both can be seen in the base of the distinctive stacks known as the Pinnacles, just west of Great Matchcombe Sand, where the schists along the shore have been eroded into isolated blocks along a diamond joint pattern. On each block is a pile of Head material which has been trimmed down more successfully by the sea and others can be seen nearby, in process of separation from the cliff.

Lannacombe beach was once the site of a watermill. The ruins are near the cottage and two solid granite millstones still lie in the sands. In the rock benches east and west of the beach quartz veins run through the schists in irregular bands. This banding is a typical feature of schists and is termed 'streaking out'.

A short walk up the valley towards Hollowcombe brings the walker up to the plateau level where the smooth surface produced by the 430-foot sea can be seen on every side. Looking north over the South Hams, there are the lower hills of the Meadfoot Beds rising in the distance to the more resistant Dartmouth Slates outcrop. In terms of what is still to be seen today, the South Hams is perhaps where Devonshire geology begins—with the old metamorphosed rocks and the lowest visible beds of the Devonian period.

CHAPTER 3

Torquay

Most towns conceal their geology only too well, but Torquay is fortunate in having a long coastline. Mid-Victorian in origin and primarily a residential town, even its continued growth has done little to obscure this natural feature and the town still retains a great deal of beautiful scenery around and within its borders. The building activity in fact increased the rock exposures as quarries seeking the local limestones were cut into parts of the coast and into the gorge walls along the Fleet Valley. Since many terraces and houses now stand in front of these old quarries, it is worth glancing round the back of buildings whenever opportunity occurs.

The geologists' most useful ally in Torquay, though, is the long coastline. The headland, a complex mass of Devonian beds, has been completely stripped of the red sandstones which surrounded it, revealing once again the limestones, shales and volcanic beds. Since each group has reacted in its own way to the work of the sea, the cliff walks are delightfully varied, their outlines and profiles constantly changing. Visitors can choose a different beach every day of the week.

Torquay is perhaps most satisfying when seen from the sea and every geologist should begin his study of it with a boat trip from the harbour to Oddicombe. Trees and greenery are everywhere above the grey-white cliffs, with the Victorian villas laid out along the contours re-emphasising the natural line of the landscape. At times, the coastal journey provides better views than those from the shore—of the raised beach at Hope's Nose or the small anticline at the base of Black Head, for example. But despite a fine coast and many quarries, the geology of Torquay has its difficulties.

Torquay stands on much-faulted and thrust structures. Thrusting, the process by which rocks are slipped over each other by low-angled faults, has repeated the rock sequence in

MAP 5

(Generalised from maps of the Institute of Geological Sciences by permission of the Director)

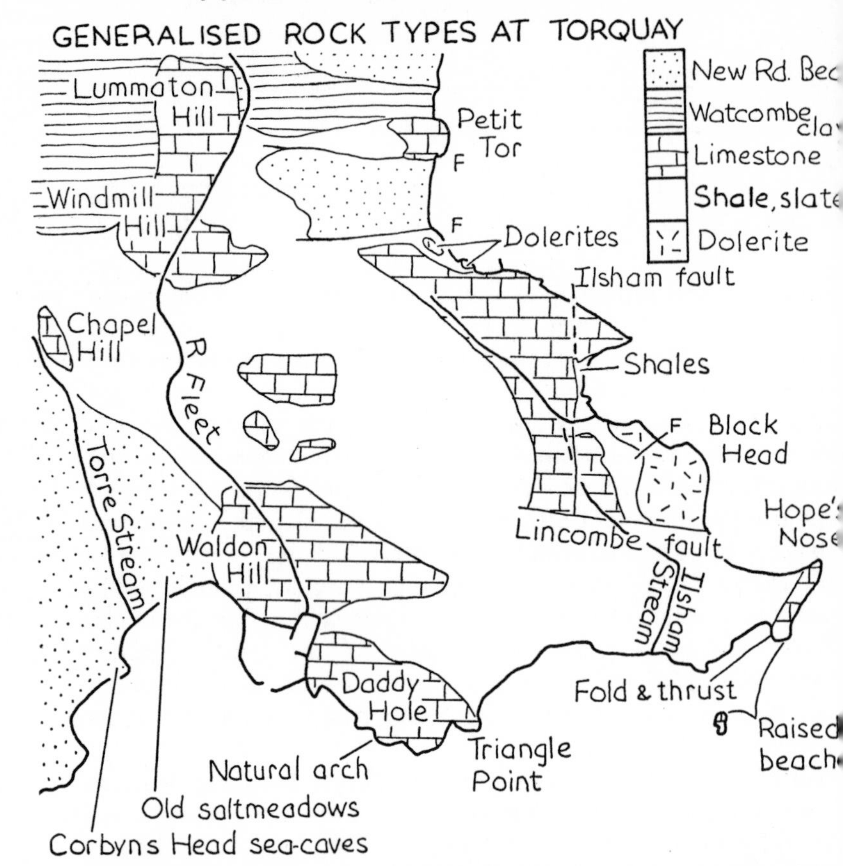

many parts of the town, and block faulting later added to the complexity.

Older grits and slates of Lower Devonian age appear in the centre forming a line of hills—Kilmorie, Oxlea and Warberry. These seem to be faulted down into Middle Devonian shales and limestones. The limestones have flat, bevelled summits produced by former sea levels. On the north, at St Marychurch, Babbacombe and Wall's Hill, the marine surfaces lie between 286 and 300 feet, but on the southern side of the headland they only reach 200 feet—at Chapel, Waldon and Park Hills, and on Daddy Hole Plain.

In theory, the rivers of a headland like Torquay should

flow directly north or south down the sides of the ridge, only disappearing when meeting a limestone outcrop. Surprisingly, this is not so. From Babbacombe, a continuous valley can be traced to near Anstey's Cove, then southwards right across the headland by Kents Cavern and Ilsham to Meadfoot beach. A similar journey across the headland was chosen by the river Fleet—and its valley reveals a fascinating relationship to the local rocks. It is an important valley for Torquay, containing the main street and several major roads in other parts of the town.

THE FLEET VALLEY

Since the Fleet is now enclosed, it is as well to say that its route lies beneath the Teignmouth and Lymington roads to Torquay Assembly Hall. It then runs beneath the south side of the main street to Cary Green and the Pavilion. Now filled with roads and houses, it was once regarded as the most picturesque valley in the district.

The Fleet rises in a broad depression to the west of Petit Tor. Formed of Watcombe Clays, this bowl is dominated by the wooded escarpment, of Oddicombe Breccias known as Watcombe Heights. Like the Watcombe Clays, the breccias are a New Red Sandstone formation, dipping gently northwards to Teignmouth. The A379 climbs steeply up the scarp at Watcombe Heights, which also serve as the site of Great Hill reservoir. Receiving water pumped from Dartmoor, the high escarpment allows a good gravity supply to most parts of the town.

The Watcombe Clays are weathered New Red Sandstones with a lot of slate fragments rather than true clays. They reach 120 feet in thickness, and formed the basis of an important terracotta pottery opened in 1867. For this type of ware the clays were regarded as the finest in England at that time and the industry produced a variety of articles, including the famous Devon mottoware. The dark red and brown body of this product was obtained by mixing the local red clay with manganese, Dorset clay and flint. A cream colour was obtained by adding ground flints and Cornish granite sand. The mixtures for glazes also included these items, with red lead, soda and borax. The traditional cream cover through which the words were scratched to reveal the dark body beneath came from dipping the ware in a 'slip' of ball clay obtained from Bovey Tracey (Chapter 12). The works ceased production in the late 1950s but its site can be seen on the A379 by Grant's marble works, and a small pottery still works similar materials at Babbacombe—standing paradoxically in a limestone quarry!

It is difficult at times to visualise the original landscape of a built-up area, and years of careful observation, recording exposures in road works and trenches, may be necessary to reveal the underlying geology. In Torquay, the problems are acute despite the long coastline. Much can be seen though from the many hilltop vantage points. So, by climbing to the top of Petit Tor, the next feature of the Fleet Valley can be appreciated. From the top, the geologist has a fine panoramic view of the marine level trimming the limestone surface from Wall's Hill to St Marychurch. West of the prominent church tower and spire the eye can trace the continuation of this

FIG 15

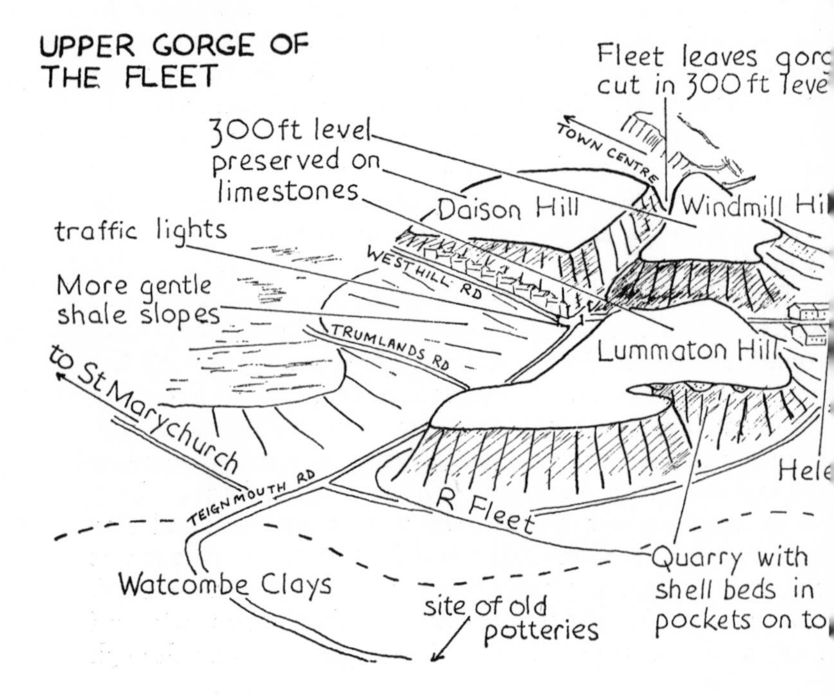

level over the Fleet Valley to Daison, Windmill and Lummaton hills. The Fleet seems to plunge directly into these limestone heights, passing in a narrow gorge followed by the Teignmouth road.

Standing at the gorge entrance, by the traffic lights at the bottom of Westhill Road, the form of the valley varies noticeably. Upstream the slopes below St Marychurch are gentle since they are formed of weak shales and ashes. The limestone slopes of Lummaton Hill on the opposite side of the valley are much steeper, and downstream Daison and Windmill hills almost meet in the narrow gorge (Figure 15).

The gorge marks the point where the Fleet cuts through the edge of the 300-foot marine levels as it flows south to the town centre. Only when it had cut through these limestone heights was it able to enlarge the softer shale and Watcombe Clay basin at the head of its valley.

LUMMATON HILL

Lummaton Hill is considerably quarried on the north side, and its grey limestones are interspersed with patches of sugary-white dolomitic limestone. Dolomite is formed by direct chemical precipitation, from waters rich in calcium and magnesium carbonate around an existing carbonate reef. The problem for the geologists is that this process involves a change in volume, destroying the structures which limestones reveal from ordinary reef accumulation. Unfortunately, bulldozers have stripped off the top beds of the Lummaton quarries, so destroying the famous shell beds, probably the most fossiliferous beds in Devon. Masses of brachiopods could be found, the bivalve shells where each side is symmetrical in itself but is not like the other side. Colonial corals like bryozoa, which became extinct by the end of the Carboniferous period, were also numerous.

The Lummaton limestones formed when coral reef conditions were well established. Elsewhere in Torquay, eg, at Dyer's quarry near the harbour, the limestones are sometimes thinner and more shaly. The latter represent early times, when the waters were still being interrupted by other sediments. Figure 17 shows the conditions in the stable reef-building times. The changes towards the left side of the

FIG 16

FOSSILS FROM THE LUMMATON LIMESTONES

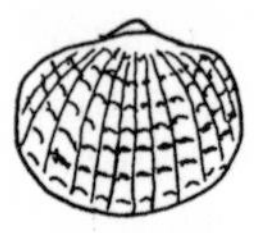
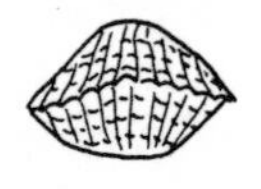

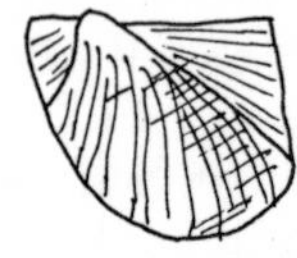

Thamnopora, Phillipsastrea and Stromatoporoids are also common – see illustration in Chapter 1

FIG 17

CONDITIONS IN THE MIDDLE DEVONIAN SEA

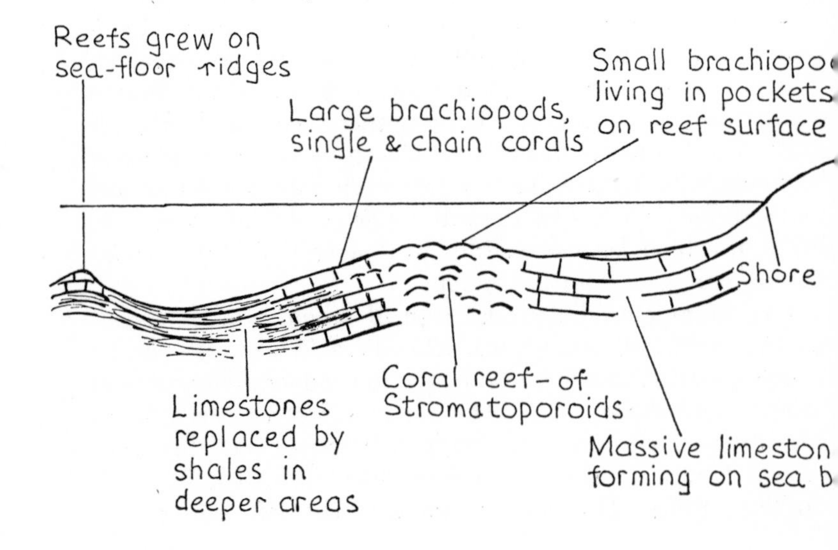

diagram are inferred and have never been observed. The reefs probably grew around volcanic islands, areas of shallow water in an otherwise deep ocean. The massive limestones they produced would have been muddy if a coastline producing normal sea-floor muds had been anywhere near at hand.

The reef life, as modern examples show, was a complex communal existence. The main structure was built up by masses of stromatoporoids and bound together by corals growing in net-like chains and colonies. Solitary (rugose) corals attached themselves in suitable spots and around them all lived numerous shell-dwelling creatures. Some of these have more heavily-ribbed shells, indicating that they lived in higher, more exposed parts of the reef. In the sheltered hollows lived thin shelled creatures like the terebratulids (Figure 16).

The northern side of Lummaton Hill is faulted, the movement bringing down the softer New Red material against the limestones and accounting for the abrupt margin of the Watcombe Clay basin at this point.

Downstream from its much-quarried gorge, the Fleet Valley opens out again, passing through shales and grits with a few small limestone outcrops until it reaches the Assembly Hall. The Jenkins marble works mentioned in the Ashburton chapter lies in this part of the valley. Much of the valley floor around Lymington Park has been filled in and the original level can now only be seen at the back of the town hall. In the main street below Castle Circus it can be traced behind the Gas Board showrooms.

At a first glance Torquay seems to be an entirely Victorian town in appearance. But step behind its main street shops and there it becomes much more of a Westcountry fishing village. Steep piles of steps and cottages perch one above the other on the much-quarried slopes between Castle Circus and the Strand. The Fleet is once again held in the jaws of a gorge, and by means of it now passes through the 200-foot marine level.

Market Street lies in a tributary gully and high above its junction with the Fleet, from the top of Braddons Hill, the gorge makes an impressive view (Figure 18). In 1800, the Fleet estuary ran right up the centre of this view to the bottom of Abbey Road, and beach deposits were dug up there

FIG 18

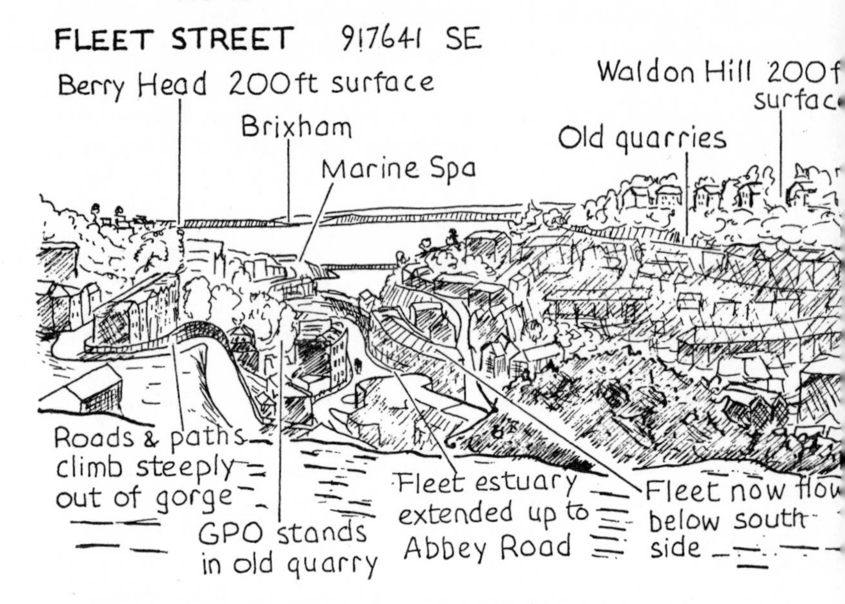

in the mid-nineteenth century. With the growth of the town, the estuary soon disappeared and the reclaimed land was extended out beyond the river mouth to form Cary Green.

Having crossed the headland by the Fleet Valley, the explorer can now retrace his steps to Petit Tor and make the journey again, this time following the coast. Petit Tor has some important quarries.

PETIT TOR MARBLE

Like Ashburton and other South Devon marbles, this is really a limestone capable of taking a good polish. It was obtained from two sites—the quarry within the knoll itself and another by Petit Tor Cove, part of the limestone which forms the point below. This lower working included the quarry known as 'Giant's Armchair' at the end of the point, as well as the shelf in the cliff face by Petit Tor Cove. It was evidently hard going on the shelf as its narrowness testifies, and very little was worked there. In the face behind it there are some interesting little cavities to be seen, lined with bladed crystals of barytes, arranged rather like bunches of leaves.

Another feature here is the way in which Upper Devonian shales have been faulted in under the limestones—near the junction of quarry shelf and beach. A very good fold is revealed in reddish shales with two cleavages (Figure 19). One cleavage runs parallel to the axis of the fold but the other is fan-shaped, arranged around the hinge of the fold. Both

FIG 19

OLD AT PETIT TOR COVE 927664E

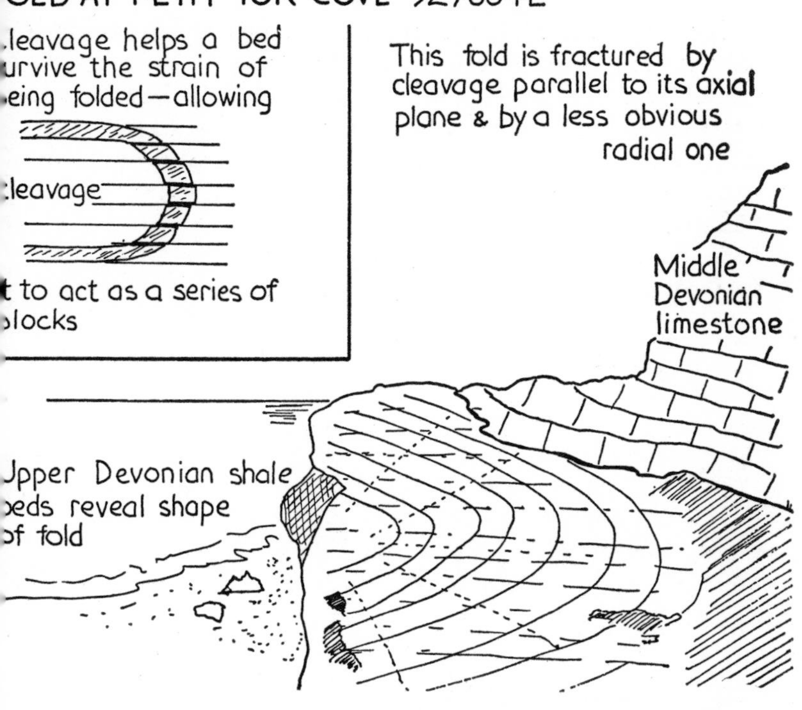

show how the rocks have tried to accommodate the pressures exerted on them. Thin, dark sandy bands help to illustrate the features.

Petit Tor marble was noticed commercially in the late eighteenth century. Stone-working families moved in from Broadhempston, and St Marychurch soon became a village of marble masons. Overmantels, pillars, steps and small items like inkstands and candlesticks were shaped up in little workshops behind Fore Street. Some of the buildings still remain behind shops on the north side while the former Hampton

Court Hotel and the large garage behind it were the home and works of one of the chief marble makers, John Woodley.

Petit Tor marble was sent to London, and even as far afield as New Zealand. A great deal of it can be seen a mile away in All Saints Church, Babbacombe, where the Gothic revival pillars and the walls lining the sanctuary are made of it.

ODDICOMBE TO ANSTEY'S COVE

The view around Lyme Bay from Babbacombe Downs is famous. The broad sweep of red sandstones which takes the eye as far as Budleigh Salterton is followed by Jurassic and Cretaceous cliffs. Continuing in a series of rather hazy humps, the view can extend to Portland—but only on rare days in the year and it takes some experience to tell whether the last hump seen is really Portland or not.

Babbacombe Downs is probably the most convincing vantage point in Devon for the junction of the old Palaeozoic

FIG 20

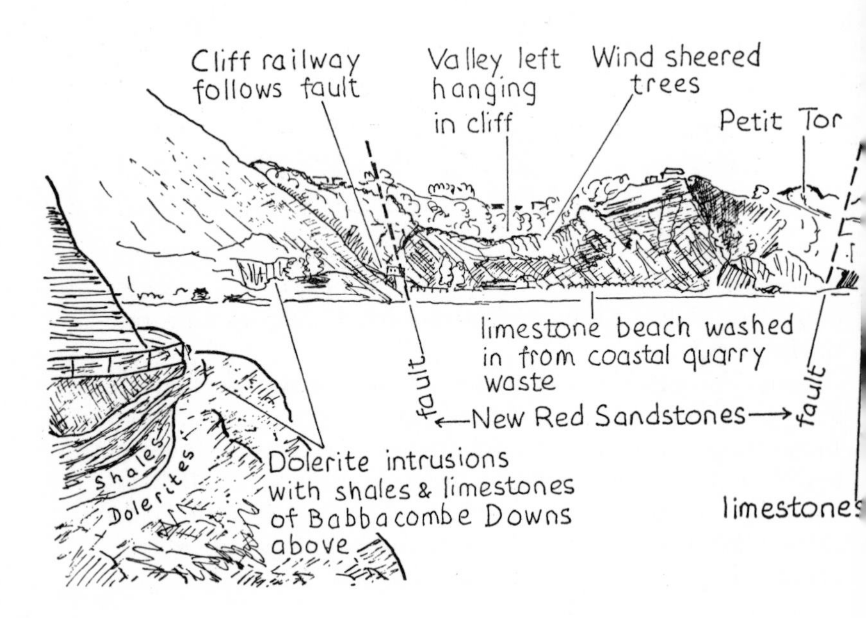

material of highland Britain with the less resistant (Mesozoic) rocks to the east.

Below at Oddicombe beach the junction has been faulted, a block of red sandstones being let down into the Devonian rather like the keystone falling from an archway. One fault bounds the south side of Petit Tor Head, the other lies beneath the cliff railway. In the sandstone block between them a little valley, now dried up and left hanging in the cliff face, makes steep gradients for walkers following the coast path (Figure 20).

The paradox of the white limestone and quartz beach below a red sandstone cliff is due to nineteenth-century quarrying further out along the Torquay headland. Boats used to tie up alongside Long Quarry below Wall's Hill and load stone for Exeter and other towns. Waste thrown into the sea was moved along to Oddicombe beach by the current, but since the quarries closed the height of the beach has dropped considerably.

The cliffs at Babbacombe Downs have a triple structure. The base is composed of dolerite sills, best seen at Half Tide Rock north-west of Babbacombe beach. The coast path passes over the top of the sill and its junction with the grey shales and mudstones above is easily traced near the seaward side of the path. The shales and mudstones are the second element in the cliff form, creating a steeply sloping section (Figure 20). Just beyond the little footbridge, goniatites like *Probeloceras* and *Tornoceras* can be found, preserved in pyrite material.

Goniatites are very valuable coiled shells with two most desirable features for the geologist—they were widespread, and they evolved and changed rapidly. So they are useful for correlating widely separated exposures and also for determining the rock sequence and age at any one locality. Here the fauna, of Lower Frasnian age, is younger than the fossils of the massive limestones at the top of the cliff. They show that the Babbacombe Down cliffs are upside down now compared to the time of their formation—a clear pointer to thrust movements carrying older beds into position over younger ones.

The base of the limestones, the topmost element of the 280-foot high cliffs, is revealed by a small spring emerging a little farther along the path from the goniatite locality. Continuing to Oddicombe beach, the top of the dolerite sill can also be seen several times.

On the eastern side of Babbacombe beach the walker following the path to Fisherman's Nose can observe a number of faults in the limestones which are here at beach level. One fault, enlarged by a stream, forms a high fissure in the face blocked off at the rear by fallen blocks. Returning to the pier, notice the intrusion of bright red felsite beside it.

FIG 21

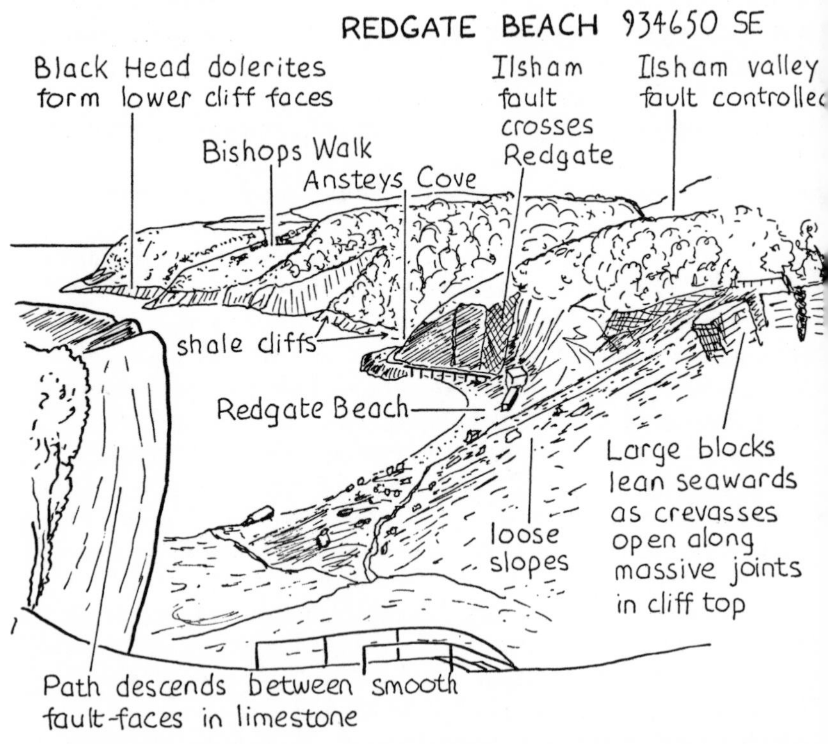

Babbacombe beach was the old fishing harbour for the houses clustered in the combe and the Victorian village which spread over the marine levels above. These levels, including St Marychurch and Plainmoor to the west, form the most extensive flat ground in Torquay and reach their eastern limit in the complex cliffs of Wall's Hill.

The best view of this cliff is from Bishop's Walk but, before crossing Redgate beach and Anstey's Cove to reach it, the geologist should visit the crevasses in the cliff top above Redgate. These have formed along joints which have gradually widened as the unstable blocks turn over towards the beach below. One crevasse was the site of a small bone cave.

From the top of the path to Redgate beach large faults can be seen behind the promenade leading to Anstey's Cove (Figure 21). These are part of the Ilsham Valley fault—the valley itself appears in the background and is the site of Kents Cavern (Chapter 4). The faults continue into the cliff below our vantage point, bringing in a wedge of shales down which the path descends. The smoothed fault faces created by the friction and heat of the movements appear as vertical walls on either side.

Across the bay below, the cliffs at Bishops Walk are noticeably lower. These are dolerite cliffs forming part of the Black Head intrusion.

HOPE'S NOSE

Massive and thin bedding, thrusting, faulting and a superb raised beach can be seen at the end of the Torquay headland. One result of the faulting is that the explorer returns again to limestone cliffs, while another is the shape of the headland. The small bays to north and south are being eroded into the fault marking the inland boundary of the limestones.

In the old quarry on the north side a thin-bedded limestone lies over a massive one. It has been thrust into this position and it is easy to see where, once in place, it slumped into hollows on the upper surface of the massive beds.

Just around the edge of the quarry, on the seaward side of the thin limestones, the little brachiopod *chonetes* can be found. If wind and tide are right the same spot provides a fine sight. Waves hitting the shore travel into the joints of the massive limestones below. One of these joints has a hole in its roof at the inner end and the water spouts from this to a height of four or five feet, exactly like a geyser.

Walking along the shore to the southern end of Hope's Nose, the cliffs are closely related to the massive and thin limestones. The massive beds form the wave platform, while the thinner beds above have been cut back farther from the shore. There is a large and much-photographed fold just before the raised beach is reached. Above it the line of the thrust between the two limestones can be seen as a straight line

FIG 22

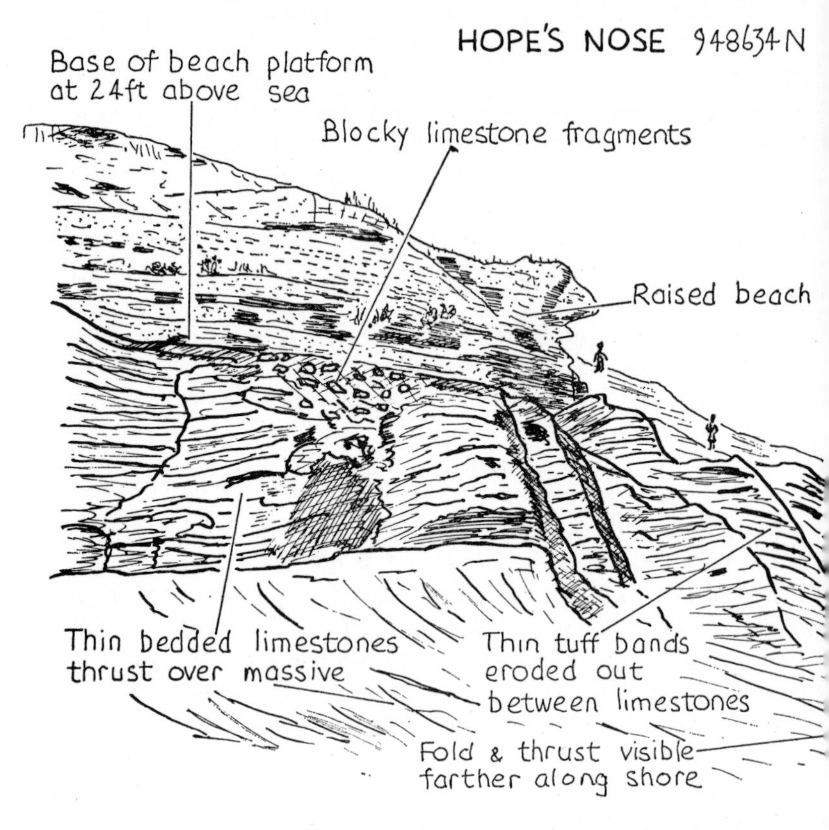

cutting across the curving limbs of the fold. The limestones in the fold are peeling off in layers, the more deeply eroded beds between them being volcanic ash.

The 24-foot raised beach at Hope's Nose is the finest in South Devon. The level can be seen at many sites but none contains such a well-preserved sample of the beach deposit as this one. From the shore, or a passing boat, the brown sand with its embedded pebbles and shell fauna is easily spotted.

A similar beach occurs on Thatcher Rock, the triangular-topped island to the south-west. Perhaps the greatest surprise about the beach is its solid construction, and securing a piece by hand is a more difficult task here than in the New Red Sandstones! It would not, of course, survive long if the sea could still attack it, but it is clearly safe for the time being with the wave force spent on the limestone platforms below. The shell fauna in the beach is a rich one—seventeen types of molluscs including *Mytilus, Ostrea, Pecten* and *Patella* have been identified, as well as pieces of crab shell. It is a useful example of recent fossil accumulation in a shore environment and dates from the latter part of the Last Interglacial phase.

Returning up the hill to the Marine Drive, fold and thrust structures can be seen again in the south-west face of the headland.

MEADFOOT BEACH AND DYER'S QUARRY

It is the frequent change of rock type which makes the coast scenery of Torquay so attractive—high limestone cliffs, lower volcanic forms, contrasts of red and white colour, numerous beaches and coves. For the geologist it also means constant changes in the age of the formations. So it is on leaving Hope's Nose. From Middle Devonian limestones he steps back to the Lower Devonian around Meadfoot beach and up to Middle Devonian again at Daddy Hole Plain and Torquay harbour.

The cliffs at the eastern end of Meadfoot beach yield poorly preserved brachiopods and also reveal channels filled with sand where rivers crossed the mud flats of the Meadfoot Beds during their formation. Thick lenses and beds of these more gingery sandstones can be picked out in the cliff. Some of the blocks on the shore appear to have ripple markings preserved in them, but seen in section the ripples coincide with the cleavage. They are structural features, not preserved beach surface features.

Along the promenade, little further detail can be seen of the Meadfoot Beds. Loose material covering the face is no longer removed now the sea road protects it and the whole cliff has been 'fossilised'.

Soaring limestone cliffs mark the western end of Meadfoot reaching the 200-foot marine level in Daddy Hole Plain and containing several faults in the Triangle Point area (Figure 23). Triangle Point, beyond the beach huts at Meadfoot, is a good locality for reef fossils although it is not an easy collecting area. The limestones seldom are, for it needs a long spell of natural air weathering to reveal their treasures. Not

FIG 23

TRIANGLE POINT 928628 E

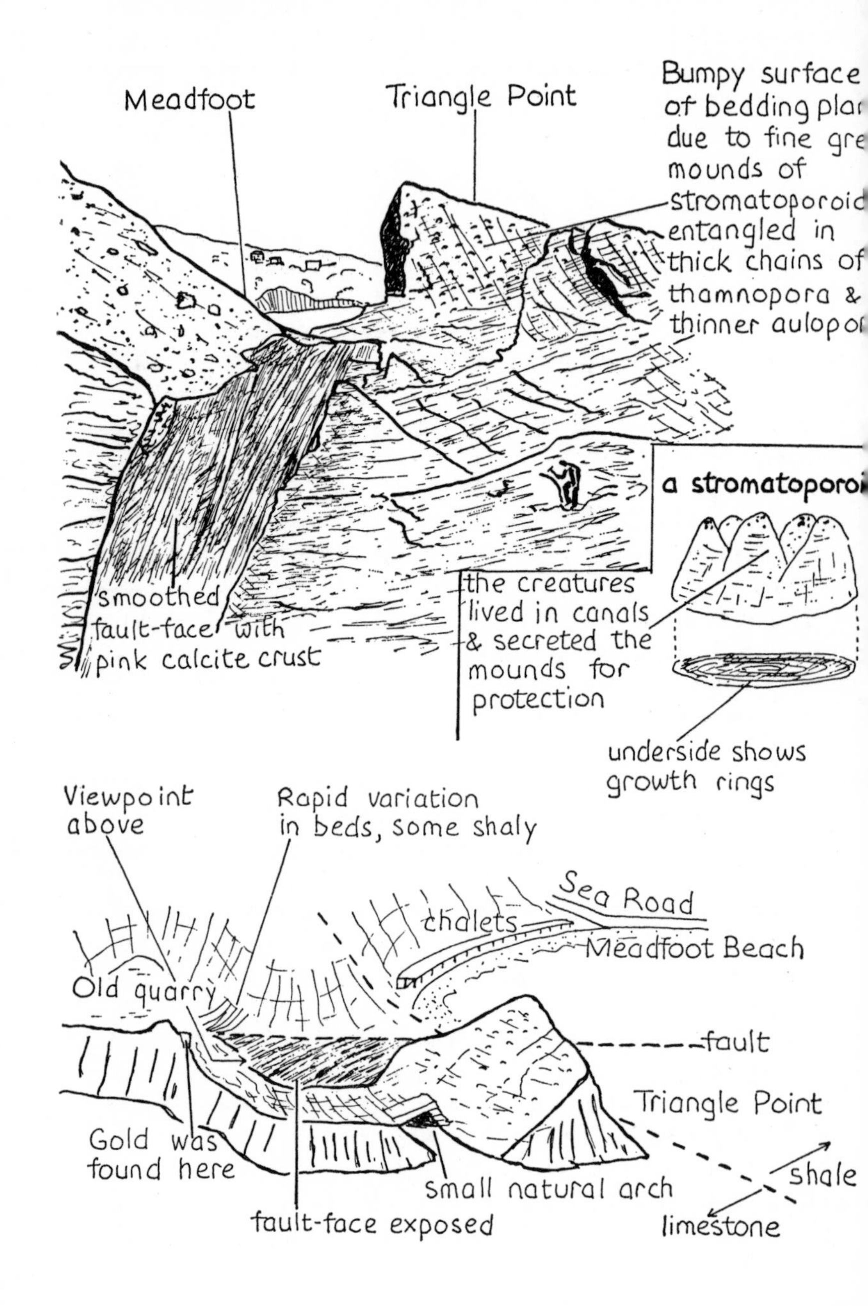

least among the attractions of Triangle Point is the fact that a small find of gold was made nearby in 1889.

The shale-limestone junction is marked by a fault immediately the end of the promenade is reached. This fault partly accounts for the shape of Triangle Point. On the steep western slope of the point the exposed bedding planes are covered in stromatoporoids and corals. The former appear as mounds and a section through one would reveal the onion-like arrangement of its layers. The animals forming these features lived in central canals, secreting the rest of the deposit as protection. The canals cannot be seen in the Triangle Point examples—what appear to be canal tubes are, in fact, other corals which grew up through the stromatoporoid structures. The stromatoporoids are linked and bound together by thick chain-like growths of tabulate corals. The thicker

FIG 24

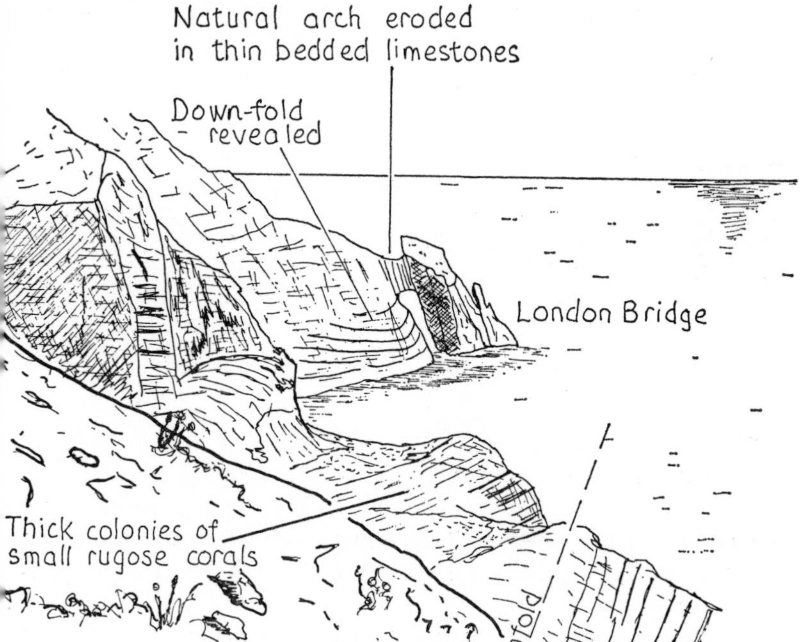

FIG 25

FOSSIL CORALS FROM DYER'S QUARRY

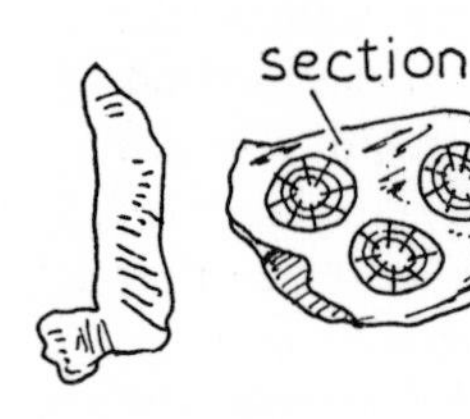

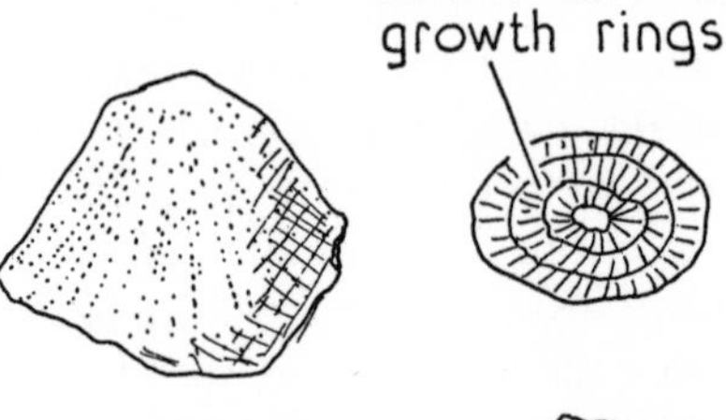

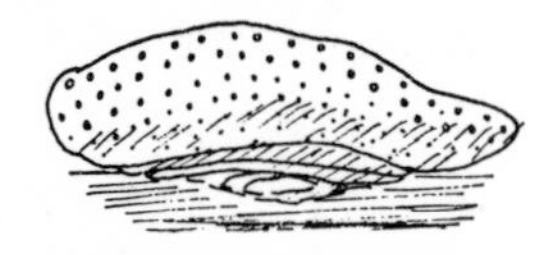

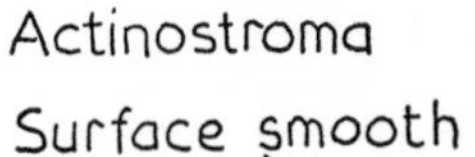

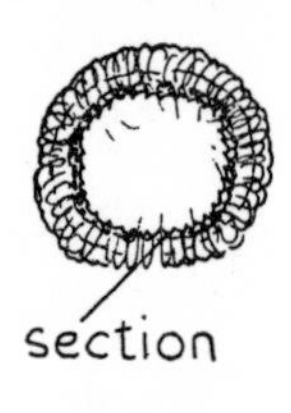

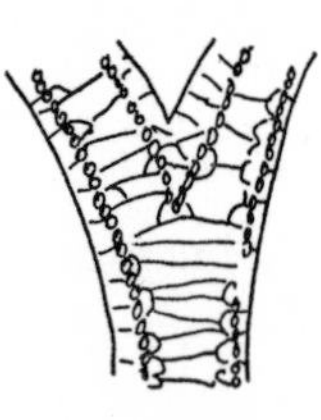

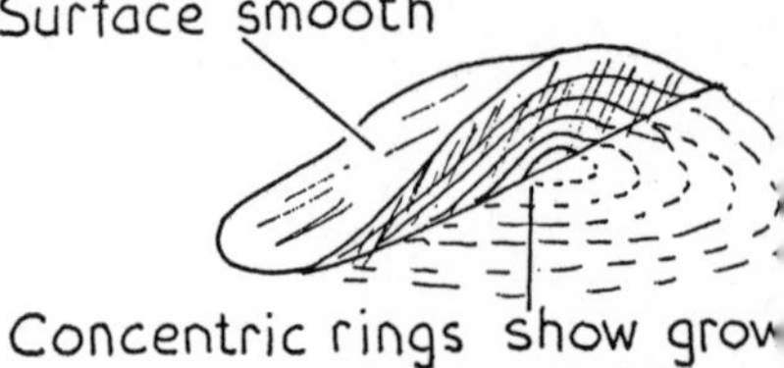

macaroni-like webs are *Thamnopora*, and among it the more whitish-coloured and smaller *Aulopora* can be seen.

Running west from the point there is an outstanding example of a fault-face. Its colour is distinctive, the limestone along it having been fused and recrystallised into pinkish-coloured calcite.

Across the face where the next quarry begins the effect of the fault on local folds can be seen, and it is in this cleft by the shore that gold was found—though not enough to start a gold rush! Alternations of massive and shaley beds in the cliff behind shows how there were interruptions of muddy conditions in the early days when the reef life was trying to establish itself here.

These earlier types of limestone beds, thinner and often pinkish-coloured, are also typical of Dyer's quarry. The quarry lies at the end of the path past the Imperial Hotel—a spot known as Lands End. London Bridge, a fine natural arch, can be viewed before going down the rather awkward scramble to the quarry. The arch reveals how the sea has removed the thinner beds behind a pillar of more massive limestones. Probably the surviving piece will fall one day, leaving only a stack formation, but its height above the sea will make this a long task.

The floor of Dyer's quarry is covered with coral faunas, many quite well weathered out of the surrounding material since the quarry went out of use. The axis of a fold can be traced across the floor, running seawards diagonally from the base of the path. On its beds are masses of small rugose corals. Although at a first glance they seem similar to the twig-like *Thamnapora* of Triangle Point, they have quite different external markings. Dyer's quarry is the best coral locality in Torquay and the most common types found there are shown in Figure 25.

ROCK WALK TO LIVERMEAD HEAD

Torquay harbour has an entirely man-made appearance and the original shore certainly had little flat land around it. The best vantage point is in Park Hill Road. The beach ran close below and followed the line of shops along the Strand to the cliffs by the Tor Bay Hotel. A rock shoal ran out beneath the present Vaughan Parade, while Cary Green beyond was then part of the Fleet estuary.

Until the turnpike era created the sea-front road, the only way to Torre Abbey was up the pathway over Waldon Hill. Rock Walk was a natural sea-cliff. At its western end the New Red Sandstones are faulted against its limestone cliffs behind the Palm Court Hotel. The line of the fault probably accounts for the trend of the Rock Walk cliff itself.

The Torre stream is the first valley encountered in the sandstones and it was the principal valley of Tor Bay before its submergence. It lies directly on the line of the Sticklepath fault and Bovey Basin and some interesting theories can be put forward in this connection (Page 167). Several tributaries joined the Torre stream as it crossed Tor Bay to its mouth, which lay off Berry Head. The old mouth is easily located in the fathom lines of the Ordnance Survey map and again the geologist can speculate on connections to the Sticklepath fault.

Beyond the former salt marshes of Torre Abbey meadows the sandstones form low headlands at Corbyn's Head and Livermead. At Corbyn's Head the bright red cliffs have been honeycombed by sea caves, while at Livermead there are some unusual features—beekites. Resembling pebbles, they have surfaces of chalcedony and calcareous cores. Beekite covered pebbles are the work of solution acting on limestone pebbles buried in the sandstones. Chalcedony, entering the cavities where they were buried, was able to build new shells around each pebble. Sometimes the core is now found to be detached from the crust—the beekite rattles when shaken. In other cases intact Devonian fossils can be found inside them, or the walker may find some with so little core left that they will actually float on water! One hotel at Livermead Head keeps a close lookout for geologists visiting the easily eroded cliffs, but beekites can also be found at Roundham Head, Paignton.

CHAPTER 4

Kents Cavern

One of the best-known show caves in the country, Kents Cavern lies in the hillside below Wellswood, Torquay, its entrance overlooking the fault-controlled Ilsham Valley. It is a vadose, or stream-eroded cavern, probably fed by swallow-holes on the surface of the limestone terrace which covers most of the valley side in this area. The way the water left the caves has never been convincingly discovered.

CAVE FORMATION

Geologists classify caves into two general groups based on the way they were formed. Caves produced below the level of ground water are termed 'phreatic'; (see Buckfastleigh, page 85). Developing in areas where the movement of water is restricted, they have a slower rate of solution and their passages show little evidence of stream action. The other group are termed 'vadose', caves produced by underground streams, and these have more complicated deposits in them since a good deal of material is brought in with the water. Two easily distinguished features of vadose caves are channelling of the roof—virtually an upside-down river bed—and notches running along the walls where the stream has widened the passage wall beside it.

Whatever their origin, the key to cave formation is the solution of limestone by water. Rainwater mixing with carbon dioxide in the air becomes weak carbonic acid and is then able to change the calcium carbonate of the limestones into a soluble form and carry it away. The solution is often redeposited in other parts of the cave in drip formations like stalactites and stalagmites, or in crystal-lined pools. These are among the most beautiful features of show caves and an

FIG 26

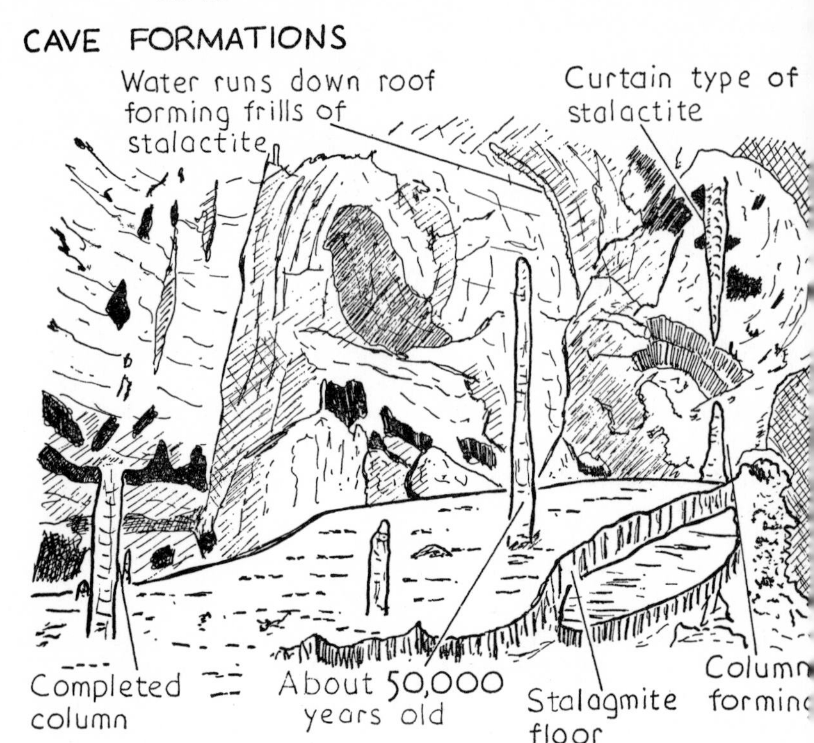

easy way to remember who's who is to say that stalactites have to 'hang on tight' to the roof.

Sometimes the passage of air currents twists the dripping water round the formation creating the strange branching helictites. The proportions of features vary, too. A slow drip means that most of the calcium carbonate is deposited beforehand and the stalactite is larger than the stalagmite below. A fast rate of drip produces the opposite result.

THE CAVE DEPOSITS

As a spectacle Kents Cavern is unsurpassed, the beauty and colour of its formations superb. For modern geologists its excitements are historic and renewed studies would probably add little to its message. The chief feature to note on a visit is the vast quantity of material which has been excavated from the system. Look for the line of the uppermost stalagmite floor high up on the walls and try to visualise the stream flowing through the cave and gradually filling it up to that level.

There were three stalagmite floors in the system, each representing a quiet period when little material was being washed in. Beneath the top crust was a black earth where items ranging from the Bronze Age to the Middle Ages were found. The middle stalagmite floor was very crystalline in character and covered a red, clayey earth with flint implements in it. It is known as the Cave Earth. Beneath it, the granular third stalagmite floor is covered by a deposit in which early animal and human remains lay side by side.

The Cave Earth is an interesting layer—the Rev MacEnery, an early excavator of the cave, termed it diluvial mud, and in keeping with the religious belief of his day he attributed it to the Deluge and the Ark. James Widger, another early excavator of Devon caves, also recorded 'diluvial' clay in his explorations at Torbryan, 1870-77. During the latter year, in

FIG 27

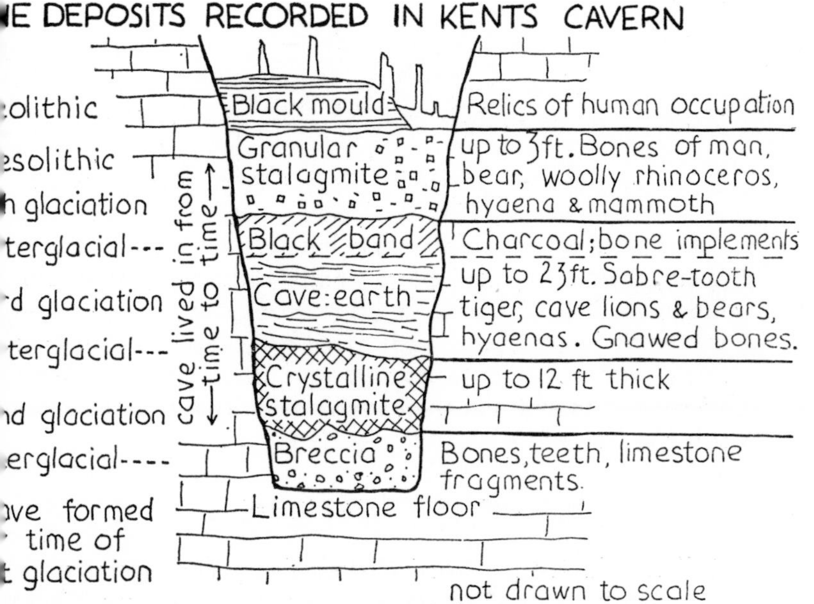

the deepest part of his digging, he left a message in a bottle—'Entered this vault for the first time on Monday, 3rd December 1877—J. L. Widger. How wonderful are thy works O God'. The bottle was rediscovered in 1954 when the Torbryan deposits were proved to be the longest cavern sequence of Ice Age beds yet discovered. They extend back beyond the Last Interglacial material of Buckfastleigh (Chapter 5) or the Cave Earth of Kents Cavern into the previous or Penultimate Interglacial phase.

FIG 28

SOME OF THE CREATURES WHOSE BONES HAVE BEEN FOUND IN KENTS CAVERN

Many of the finds made in Kents Cavern can be seen in the Torquay Natural History Society's Museum in Torwood Road. The bones included many long extinct animals: cave hyaena, mammoth, cave bear, woolly rhinoceros, cave lion, sabre-toothed tiger, grizzly bear, bison, reindeer, Irish elk and horse were found (Figure 28). The best way to familiarise oneself with the vast literature and importance of Kents Cavern is to tour it and then study the written records before visiting it again.

THE ANTIQUITY OF MAN

If Kents Cavern is a 'passive' study for modern geologists, it was certainly nothing of the sort in the last century. Those days saw the peak of its geological excitement.

FIG 29

KENTS CAVERN & NEIGHBOURING LIMESTONE OUTCROPS

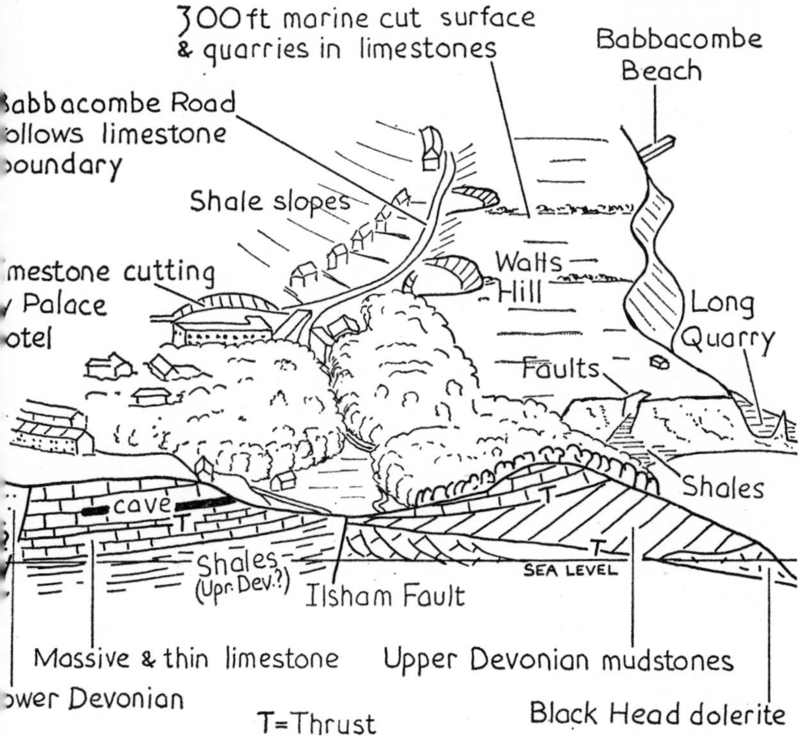

In the early nineteenth century the popular religious belief was that man and his world had been created one morning in the year 4004 BC. Every new fact that came to light had to be fitted into the Biblical time scale, and many good scientists consequently missed the full importance of their discoveries. The all-important fact was the Deluge—everything was diluvial or antediluvial, including the vast clay drifts covering much of England. Man, it was said, was not as old as the Ice Age and its extinct mammals, so their bones were believed to have been washed into the Cave Earth by the Deluge itself. All these issues were eventually resolved in the momentous years 1859-60. Kents Cavern and the studies of three men—the Rev MacEnery, Dean William Buckland and William Pengelly—led to the discoveries of the century.

No bones were discovered in Kents Cavern until 1824, although Dean Buckland had visited the cave previously. From 1825 to 1829 MacEnery, a local priest, excavated in the cave and found flint implements lying next to the bones of extinct mammals. A man of poor health, wrestling with the conflict between his own evidence and established belief, MacEnery was easily rebuffed by the great Buckland's refusal to support his findings. Buckland preferred to explain the implements as left in hearths dug into the floor of the cave. He visualised early man making his fireplaces and inadvertently mixing his implements with the bone-bearing Cave Earth a few feet below. MacEnery knew there was no evidence of this but could not contradict the greater authority who, only two years earlier in 1823, had published his *Reliquiae Diluvianae* and was the first professor of geology at Oxford.

Dean Buckland must not be thought of as an immovable man however. He could admit himself wrong and in 1838, for example, he readily championed recognition of the great clay drifts as glacial deposits. Needless to say, some of his fellow churchmen did not regard his views too happily! No doubt he would eventually have accepted the antiquity of man, had he lived long enough.

As it was, a second period of excavation in 1846 repeated MacEnery's proofs. Conducted by Edward Vivian and William Pengelly, its report was withheld by the Geological Society until Buckland could visit the cave again. Unfortunately, he never had the chance to do so.

In the heat of all these arguments the virgin cavern at Windmill Hill, Brixham (Page 102) was discovered and excavated in 1859—and so scientifically that the matter was finally placed beyond doubt. Pengelly acted as local agent for Dr Falconer of the Geological Society of London, which was responsible for the work. Fresh studies were then made at Kents Cavern between 1864 and 1880. Palaeolithic man had definitely existed in the Pleistocene (Ice) Age, a contemporary of the mammoth and other extinct mammals.

Of the three men principally concerned, MacEnery must be regarded as blameless. As an amateur (in the best sense of the word) he was unlikely to publish such startling evidence against superior advice. Buckland, too, must have his just credit. He did recognise the difference between cave deposits of three types—pitfalls, inhabited caves, and water-laid deposits. He was an early exponent of scientific methods but died before the period of Darwin and Boucher de Perthes, makers of the scientific revolution of 1858-60. What momentous years geology went through in the last century!

With the passing of MacEnery in 1841, of Buckland in 1856, and the illness of Dr Falconer of the Geological Society, it became Pengelly's task to report the final proof. A nationally much-respected geologist and a father-figure in Devonshire geology, it is very appropriate that his name should have been taken by the cave-study centre at Buckfastleigh (Chapter 5).

CHAPTER 5

Ashburton and Buckfastleigh

The A38 road running south-west from Bickington to Ashburton and Buckfastleigh passes through an undulating landscape of limestone and volcanic rocks. Most early travellers along this route, preoccupied with the difficulties of the roads, did little more than note the local scenery. W. G. Maton in his travels of 1794-6 found 'the scenery of this part of England altogether rich and interesting'. Maton is the first person known to have attempted a geological map of south-west England. Others, like Dean Jeremiah Milles and Polwhele, made more detailed observations, particularly on the limestone. Milles wrote in 1760 'in Mr Furnese's quarry are a great many stalactites and stalagmites of a beautiful kind of alabaster, the rock being full of small cavities in which they are found though none of them go far underground'. In his *History of Devonshire,* 1797, Polwhele described the area around Buckfast Abbey as 'one continued lime-rock, which is worked at many places to a depth, height and extent surprising and forming a vast cavern at once terrific and beautiful, which proves an inexhaustible fund of gain to the owner'.

Quarries like Bullycleaves at Buckfastleigh had already been worked intermittently for centuries so their size was to be expected, and Polwhele's only error was to describe the limestones as continuous. Compared to the Brixham and Plymouth outcrops, they are found in many small patches in this part of Devon. The local volcanic rocks with their ashes and volcanic bombs also appear in numerous small outcrops while the whole zone, away from the main road, is surrounded by Upper Devonian slates. A short distance to the north-west the slates give way to Carboniferous shales and grits. Beyond Ausewell Rocks and Hole Chase, these have been metamorphosed in a zone about a mile and a quarter wide which surrounds the Dartmoor granite. Mineral deposits and other features due to the granite reach out well beyond this immediate altered zone into the limestones and other rocks at Buckfastleigh and Ashburton.

The geological fascination of this part of Devon lies near the towns of Ashburton and Buckfastleigh. Too early in history to record it, the early travellers unsuspectingly passed through an area embracing changing sea-levels, coral reefs, animal life typical of modern East Africa, umber deposits and marble as features of its geological story.

FIG 30

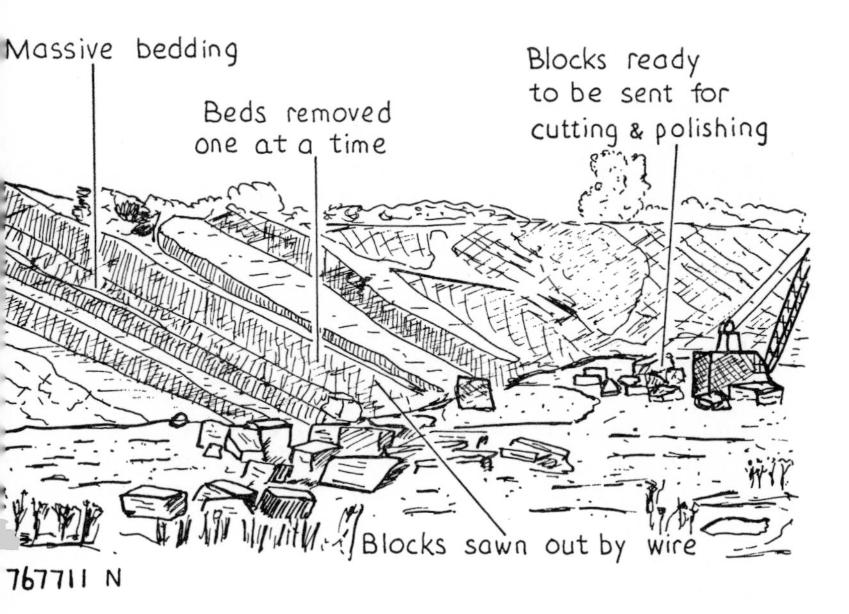

STRAWBUSHES QUARRY – ASHBURTON MARBLE

ASHBURTON

The two most important geological products at Ashburton have been umber and marble, both derived from the outcrops of limestone which surround the town.

Ashburton marble was quarried in the eighteenth century and sent to Plymouth and London among other places. In hardness and variety it was thought the equal of foreign marbles. Today, it is the only marble produced in Britain and is sent all over the world—America, Cape Town, Hong Kong.

Strawbushes quarry contains 15-20 acres of fossiliferous limestone beds, good enough to take a hard polish. Their compact nature is not due to metamorphic heating and fusing of the rock, but to recrystallisation during its folding. This fused material in the joints, producing a variety of coloured veins. So the rock is not a true marble, only earning the name from its ability to take the high polish which reveals this rich colouring. This is why it is best used for internal wall linings and paving. Outdoors, it weathers too quickly compared with the true metamorphosed marbles found in Italy and Belgium.

The varied patterns in the dark grey material come from the clay beds in the limestone and from the dense colonies of coral and stromatoporoid remains. Other colours are due to traces of iron in the rocks (compare the specular haematite in Buckfastleigh caves, below). The red haematite veins may have been washed in from New Red Sandstone rocks which once covered this area and rested on Dartmoor beyond. Calcite formations trace pure white veins across the marble and there are also bands of soapy-feeling material. These were produced by water washing through the joints and re-depositing calcium carbonate in the same fashion as when it forms stalactites.

Strawbushes quarry is not worked by explosives—the rock is too valuable. Instead cutting is done by wires, which serve as guide lines for an abrasive mixture of sand and water. The sand comes from Dorset's Chesil Beach and is the hardest quartzite material to be found in this country. It may take three weeks to cut out a 60-ton slab and there is no sign of the quarry running out.

The marble is polished at Jenkins marble works in Torquay and when demand falls, or stocks at the works are sufficient for the time, production at the quarry is eased off or stopped. The output averages between 5,000 and 6,000 cubic feet per year, and is sent mainly to London where it can be seen in the foyer of the GPO tower, in the Hilton Hotel's 500 bathrooms and in many other buildings. Many slabs are also sent uncut to the United States via Avonmouth docks. Oddly enough, it is known over there as Renfrew marble, and again features in many public buildings including the President Roosevelt memorial. Locally, this beautiful stone has featured

MAP 6

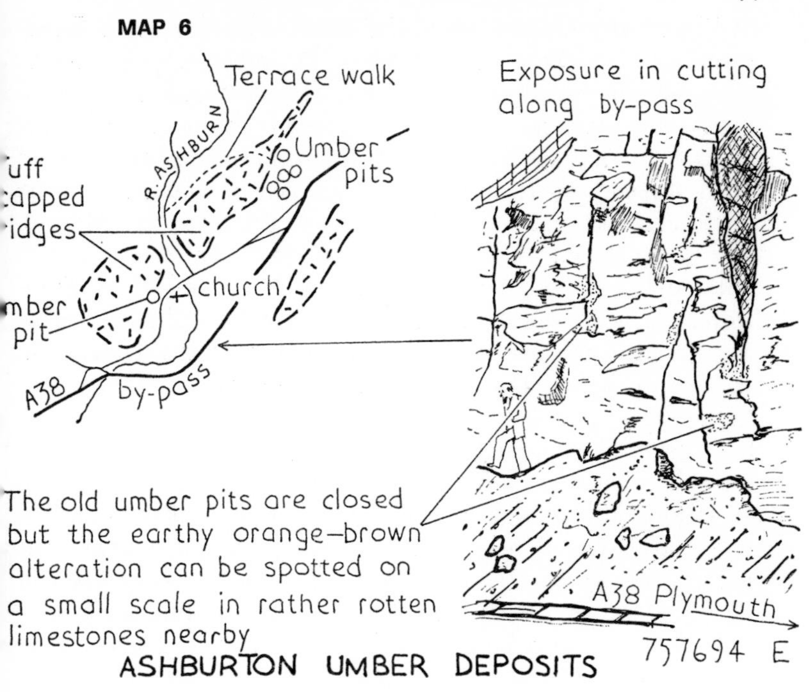

ASHBURTON UMBER DEPOSITS

FIG 31

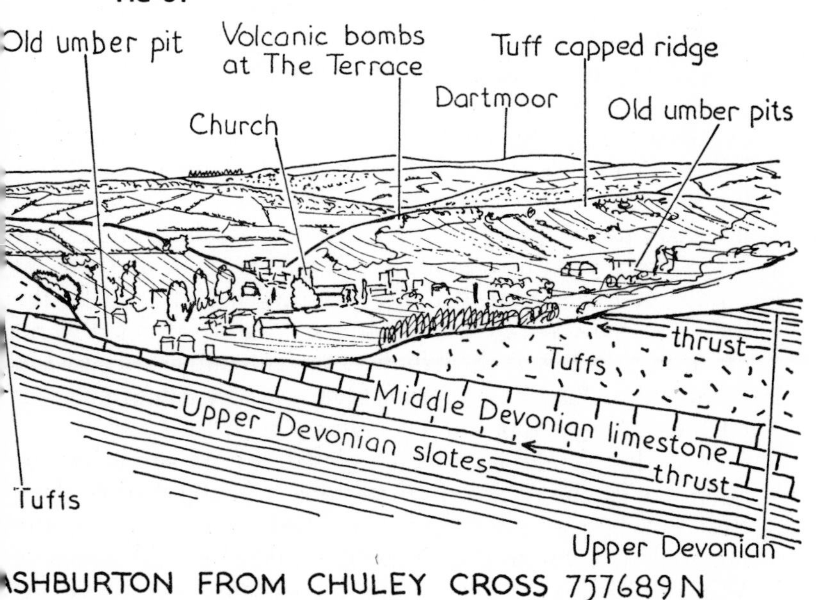

ASHBURTON FROM CHULEY CROSS 757689 N

in Plymouth's Civic Centre, Guildhall and Library, in Barnstaple's Civic Centre and in Buckfast Abbey. Other modern uses include furniture items such as coffee tables.

In the nineteenth century Ashburton was well-known for another product derived from its limestones—umber, a brown form of haematite (iron ore). This was dug at Goodstone to the north-east and from large pits close to the town (see Map 6). All the umber pits were closely related to the volcanic material outcropping on the hillsides above. The lower beds of limestone near the town are of dolomite character and water washing into them from the volcanic materials caused chemical changes to occur. The ordinary limestone of the district contains about 46 per cent lime and 41 per cent carbonate of magnesia. The chemical action dissolved away nearly all this material, leaving umber deposits 20-30 feet thick. Original variations in the character of the limestone probably explain irregularly-shaped masses left unaltered in the midst of the umber. Strings and veins of impure iron and manganese can be found running through volcanic beds and umber alike.

Umber has also been recorded elsewhere in Devon, near Combe Martin and at Broadway, near Torbryan. At Ashburton, the principal pits were the Roborough works and the Devon and Cornwall works on the north-east of the town near the Cottage Hospital. Another existed by the Bowden Hill/West Street junction, immediately west of the parish church.

After the overburden was removed, the umber was worked by digging large rectangular pits, shored up with timber sides. When the unaltered rock below was reached, the pit was filled in and another commenced alongside. The only processes necessary for the gritty, brown earth were washing and crushing. Stamps were used as in dressing metal ores, and then as liquid mud the washed umber was settled and dried by the same method used in the old china-clay drys—passing warm air beneath a tile-lined shallow tank. The umber was used in paints, in making brown paper and even as a colouring agent for the poorer types of woollen cloths.

Now that the umber pits are closed, a good site to see the alteration of the Ashburton limestone is in the cutting along the A38 by-pass. Its joints are filled with brown earthy material and volcanic rocks outcrop above on the slopes of Whistley Hill. Here and in similar volcanic material at the

Terrace, north-east of the town, volcanic bombs can be found—lumps of molten material thrown through the air during eruptions. Their distinctive shape is due to the gyrating motion shaping the still plastic material in flight.

BULLYCLEAVES QUARRY

The river Yeo, or Ashburn, drains the Ashburton area southwards to the Dart at Buckfastleigh and the A38 follows its western bank. Near Dart Bridge is Bullycleaves quarry, one of a group which almost surround Holy Trinity church on the hilltop above. The massive grey limestones here show no signs of the chemical changes seen at Ashburton.

Bullycleaves is probably the oldest quarry in the district, and offered the nearest good stone to Buckfast Abbey, where extensive building must have started after its foundation in 1018. About 1150 further enlargements by Cistercian monks

FIG 32

BUCKFASTLEIGH 749670 SW

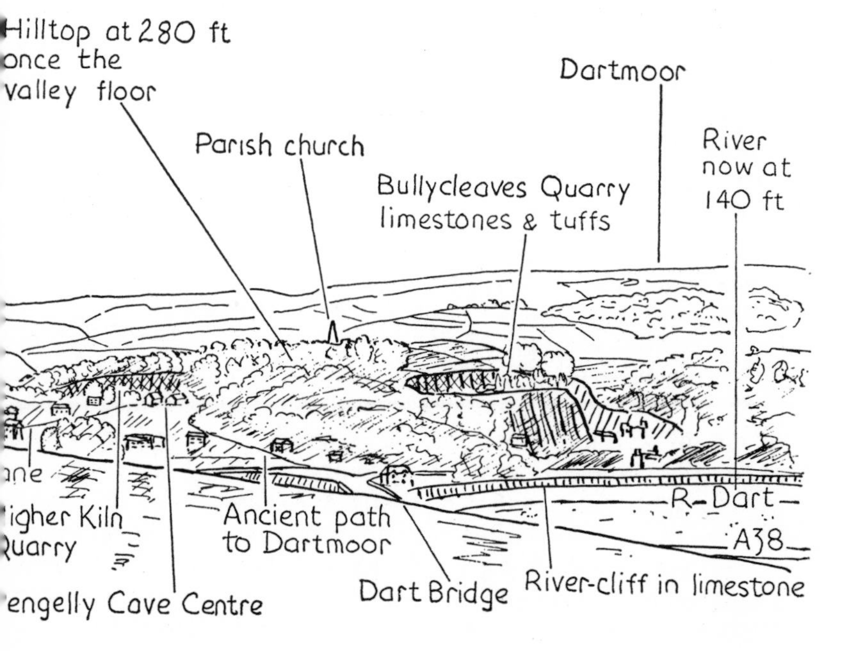

would have renewed demands on the quarry. At the Dissolution the abbey fell into ruins and over the years many local people came to take away the ready-cut stone lying there rather than work new stone from the quarry. Carved stones can be seen in many Buckfastleigh cottages today. The present abbey was built by French Benedictine monks after they bought the site in 1882.

Ancient maps describe Bullycleaves as the quarry for 'the kyng's tennents to bill there howsys and marlee there grounde to bring forth corne'. In those days they probably had little use for the outcrop of volcanic tuff in the base of the quarry but this is now a source of road gravel. On the south side of the town Whitecleaves quarry, which is entirely in volcanic material, produces 10,000 tons per year for tarmac purposes. In a limestone quarry like Bullycleaves, however, chippings are less important than other products since they become slippery with wear and so are not permitted in main road tarmacs.

THE INFLUENCE OF DARTMOOR

Both Ashburton and Buckfastleigh have been mining parishes and, apart from the tin and copper lodes once worked, there are grey and glittering, specular haematite veins to be seen all over the area in the limestones and other rocks. Several of the caves at Buckfastleigh contain these veins.

The most valuable local mines were Owlacombe and Wheal Emma. Owlacombe, two miles north-west of Ashburton, was working in this century, producing tin and arsenic. Its lodes ran east-west, like those of Wheal Emma, and the common direction of tin and copper lodes.

The Wheal Emma site is now in private woodland at Brookwood, west of Buckfastleigh. It was working in the 1861-82 period when it produced over 30,000 tons of copper. Some idea of its importance can be gauged by the nine-mile leat constructed to it from the Swincombe Valley via the Dart and Fernworthy. A number of shafts were dug down the valley side of the river Mardle, but the most interesting feature here now is a stream flowing through the old workings. It comes out so charged with copper solution at the valley bottom that twigs and stones in its bed are quickly coated green.

Climbing the steep path from Dart Bridge to Buckfastleigh parish church, a further example of the influence of the moor occurs near the gate into Higher Kiln quarry. Just inside this entrance to the Pengelly Cave Centre (occupiers of the quarry) is a dyke of lamprophyre. It was injected into the limestones at the time Dartmoor was forming, or during

FIG 33

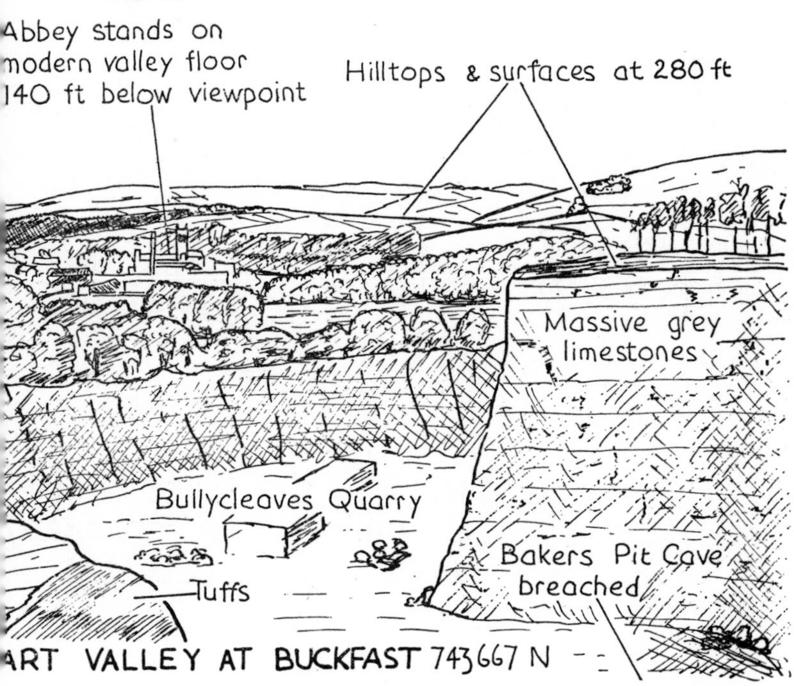

ART VALLEY AT BUCKFAST 743667 N

some later upheaval associated with it. The dyke recrystallised the limestone around it, which now appears as large pink/white calcite crystals instead of the usual grey colour. This may be the reason why quarrymen left the projecting ridge which marks the main quarry face opposite, the altered limestone being unsuitable for lime-burning. The dyke can also be traced across the Dart Bridge path below the quarry gate.

Today, the most important link with Dartmoor is the Dart itself. Its role can best be seen by continuing up the path and entering the churchyard where there are good views northwards up the valley and westward to the moor.

MAP 7

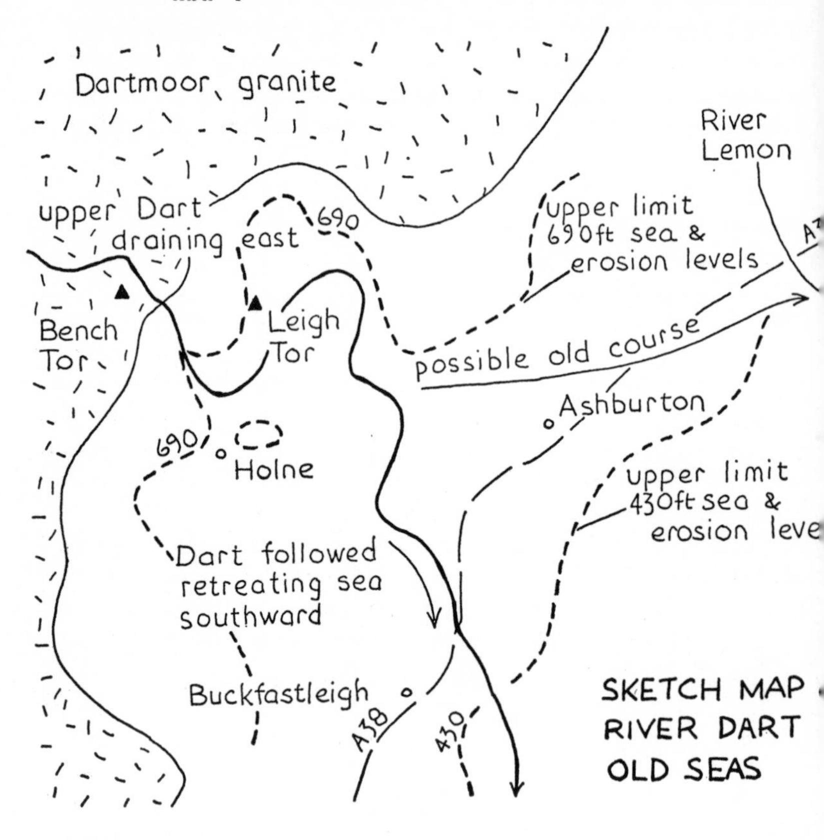

THE RIVER DART

Aided by man and his quarries, the Dart has woven together the rich story which was still undiscovered in the days of the early travellers and their journals. Its valley, with its varied beauty, is still a mecca for the modern tourist, although the estuary and the moorland parts seem more popular than Buckfastleigh.

The Dart breaks out from the granite country through the spectacular reach from Dartmeet to New Bridge. North of Bench Tor it passes the 700-foot contour and enters the metamorphosed Culm rocks running south-east in a straight, deep cleft. From this point to its mouth the history and scenery of the river have been intimately connected with the high sea levels of the early Ice Age (Chapter 2). The highest level which could have been marine in origin, is 690 feet, but there are remnants of older erosion surfaces on Dartmoor, produced under land conditions and now found at 800 and 1,000 feet, for example. In those higher areas the Dart flows eastwards, but below 690 feet its direction is southward (Map 7) and it seems to have changed direction then to follow the retreating seas. Some geologists believe that the Dart still continued to flow east after the 690-foot surface was formed, via the river Lemon and Newton Abbot, and for them it did not make the southward change of course until the seas stood at 430 feet. If so, the eastward flowing Dart could have cut the broad valley which the A38 follows towards Ashburton and if river gravels could be found there with granite pebbles in them it would be proof that the Dart once crossed the area. Unfortunately, no such gravel has been discovered and this remains an unsolved problem of the Dart's history.

All geologists agree that by the time the 430-foot erosion surface was forming the Dart must have changed course to flow south to Buckfastleigh. There, just west of the hilltop churchyard, is an earthy deposit with pebbles of waterworn granites containing tourmaline. These prove that a river coming from Dartmoor once crossed the top of this hill. The summit, now at 280 feet, was part of one of its old valley floors.

Looking up valley (Figure 33) other remnants of the Dart's sequence of levels can be seen immediately behind and above the tower of Buckfast Abbey, but none of them provides more convincing evidence than this hill at Buckfastleigh. Most of the hill is limestone—an excellent rock for preserving old erosion levels.

At the next fall in sea level, when the Dart began yet another phase of down-cutting in pursuit, the water inside the hill was quickly drained away through the natural joints. From then on streams were unable to survive on its surface or to destroy the evidence the hilltop represents. Important consequences of this situation can be seen by returning down the Dart Bridge path to Higher Kiln quarry to visit the Pengelly Cave Centre.

HIGHER KILN QUARRY: EARLY EXPLORERS

With its five caves, this is probably the quarry referred to by Dean Milles and Polwhele. The Rev MacEnery and William Pengelly, major excavators at Kents Cavern, Torquay, both visited the Buckfastleigh area which Pengelly described as a 'metropolis of caverns'. MacEnery made an unsuccessful search for bone deposits there in the early nineteenth century. Stone-working at Higher Kiln quarry had, in fact, ended some seventy years before two major discoveries were made in 1939.

In May that year new and beautiful extensions to Reed's Cave were entered and in June the bone deposit in Joint Mitnor was unearthed. Two years of digging produced over 4,000 bones and teeth from a fauna of mammals living in the last warm interglacial period of the Ice Age. The climate then was rather like East Africa today, the remains including hyaena, hippopotamus and rhinoceros. Most of these bones are now in the Torquay Natural History Society's museum. The next significant study was J. F. N. Green's paper on the river

FIG 34

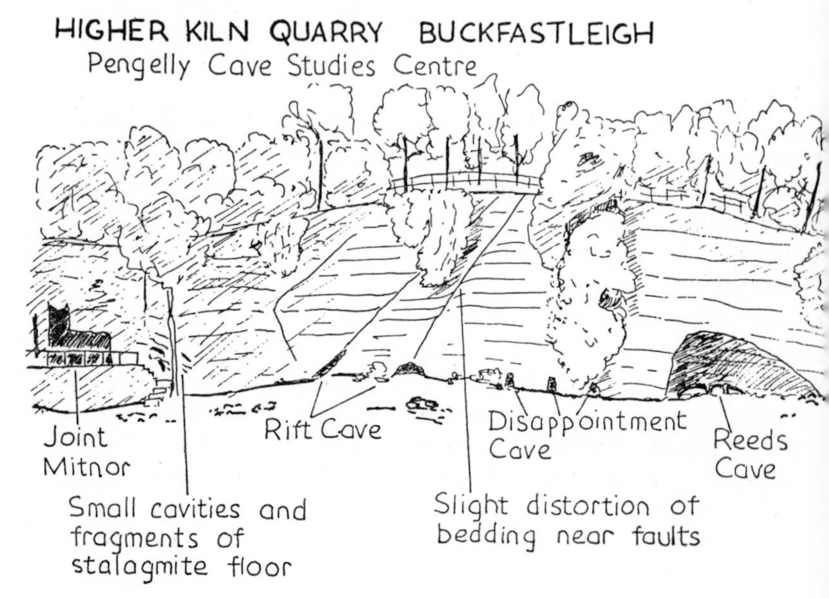

Dart, published in 1949. This work dated the hilltop terrace and the present river level and fixed the period of time during which drainage of the caves and accumulation of the bone deposits occurred.

Thus Higher Kiln quarry was of known scientific value and about to be designated by the Nature Conservancy when it was put up for auction in 1961. Its preservation was assured when the Society for the Promotion of Nature Reserves became its new owners. Through leases made by the society this quarry, its five caves, and two old farm buildings reconstructed as a museum and lecture hall, now form England's first cave studies centre. The site was not a good buy from a potholer's viewpoint, nor is it a typical British caving area. Most British caves are on open moor or commons but in Devon the majority are in private farmland and access is controlled. So, appropriately, valuable cave remains occur in the best area of the country for their continued conservation. A good deal of unintentional damage can be done to these underground wonderlands and the main purpose of the centre is to teach the proper care and study of caves. It is not a show-cave centre but caters for individuals and groups on a voluntary basis. Written application to visit must be made to the head warden.

THE PENGELLY CAVE STUDIES CENTRE

Limestone is a porous rock because of its well-developed joints through which water quickly drains to lower levels but, surprisingly, inside the gate of Higher Kiln quarry there is a pond. The volcanic ash or tuff beds frequently found in the limestones form a local impervious layer here, preventing the pond being drained away underground. The Centre's cave system also seems to be related to beds of tuff which have restricted the cavities to the upper part of the hill. The whole system was formed under phreatic conditions (see Chapter 4) by solution within groundwater-filled rocks. At that time the Dart was still crossing the hilltop and the cavities were well beneath its valley floor. When the Dart cut down to lower levels the groundwater still within the hill had to migrate laterally over the tuff beds below, briefly emerging as springs wherever the base of the limestone out-cropped in the new valley sides. The springs ceased as soon as the emptying of the cavities was complete.

Three of the five caves in the main quarry face have been developed for special purposes—a type of presentation already popular in America. Joint Mitnor cave features the bone deposits; Rift Cave is reserved for biological studies, mainly of bats; and Reed's Cave is noted for its beautiful formations.

Outside the Centre, Baker's Pit and Pridhamsleigh caves are used for exploration—a new extension to the Pridhamsleigh system was made in 1970 by divers swimming through

FIG 35

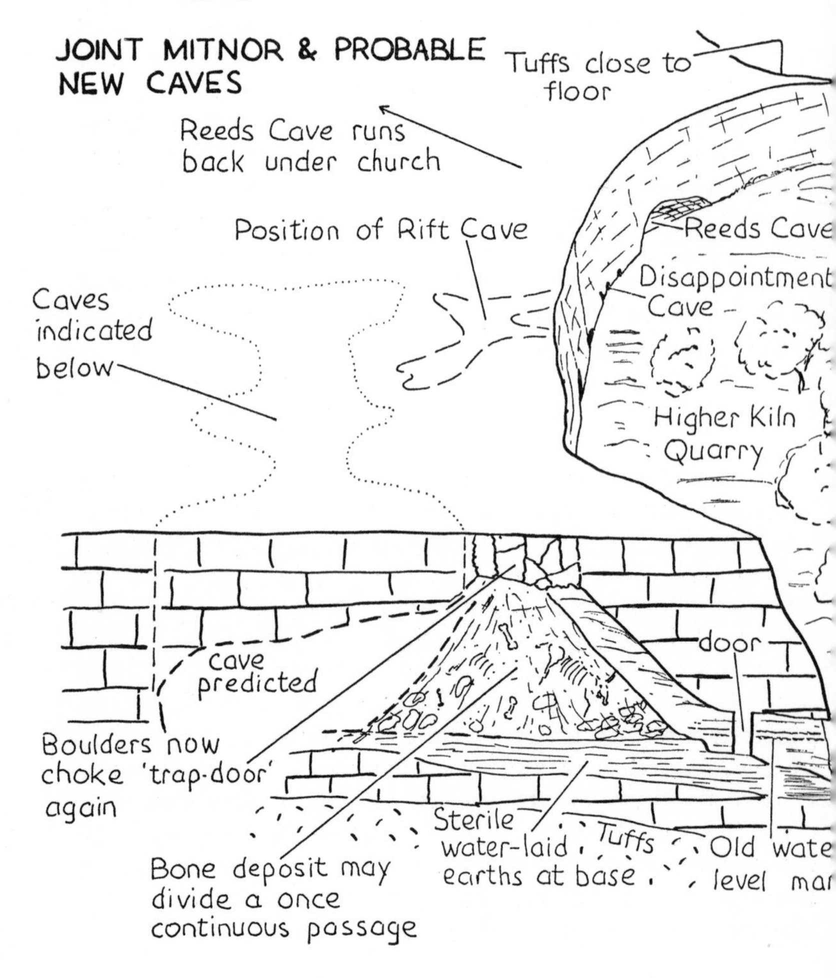

submerged passages beneath an underground lake. Baker's Pit is entered from the quarry west of the church, a quarry now used as a refuse dump, and like Reed's Cave it passes beneath Holy Trinity church. Few features of value remain in Baker's Pit, although its entrance is being preserved. The cave was opened in 1847 and Wm Pengelly visited it in 1859. It contains a pillar of specular haematite, iron ore with so much mica in it that it displays thousands of glittering facets. Baker's Pit extends right down below the church to the face of Bullycleaves quarry and when the quarry wantonly broke into it, it destroyed the delicate balance of conditions in the cave by introducing air currents and foreign forms of life to its lowest areas.

Reed's Cave lies between the Baker's Pit system and the church above. Its huge canopy in Higher Kiln quarry still shows many stalactite formations and this once complete chamber was probably the show cave referred to by Polwhele. The present entrance is high up in its inner wall. A journey through the system must be made with an experienced leader and involves a two-hour crawl, ending beneath the church at an unusual shaped pillar called 'The Little Man'. Aesthetically-minded visitors may dislike the old bedsteads built into the cave gateway but they are an effective means of letting bats in and keeping inexperienced humans out! South of Reed's Cave, one or two almost overgrown entrances mark Disappointment Cave, aptly named since they lead nowhere and the explorer is no sooner in one way than out at the other.

A large fault running down the quarry face is the reason for the formation and name of Rift Cave. Entry to it is seldom allowed as it is inhabited by a horseshoe bat colony. At the inner end of the cave is a small pool.

JOINT MITNOR CAVE

With its bone remains, paving and electric lighting Joint Mitnor is the most accessible and exciting cave for visitors without special caving equipment. It, too, has a large entrance partly quarried away like Reed's Cave, and fragments of formations and stalagmite floors outside and in the bed of the quarry opposite show that it once crossed the worked area, and probably continued under the nearby house 'Russets'.

FIG 36

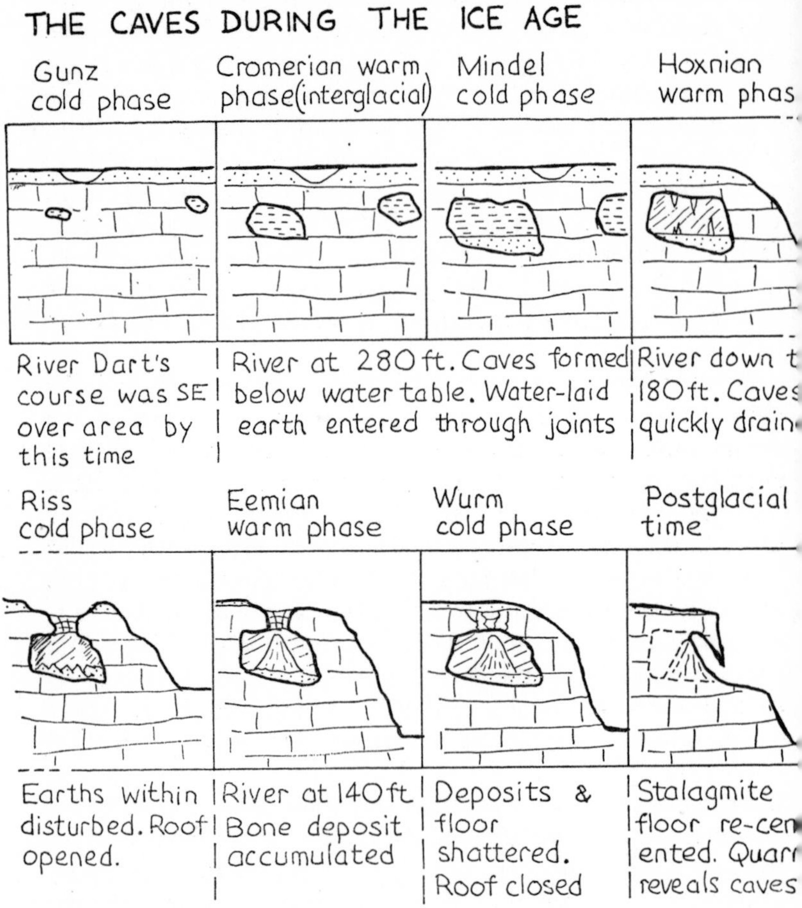

THE CAVES DURING THE ICE AGE

Gunz cold phase
Cromerian warm phase (interglacial)
Mindel cold phase
Hoxnian warm phas

River Dart's course was SE over area by this time
River at 280 ft. Caves formed below water table. Water-laid earth entered through joints
River down t 180 ft. Caves quickly drain

Riss cold phase
Eemian warm phase
Wurm cold phase
Postglacial time

Earths within disturbed. Roof opened.
River at 140 ft Bone deposit accumulated
Deposits & floor shattered. Roof closed
Stalagmite floor re-cem ented. Quarr reveals caves

Outside the door is a section of water-laid earth. As this entered along joints and accumulated within the cave while it was still filled with water, it contains no bone remains. On the right hand wall over the entrance are traces of two old water levels. They can be traced into the cave and indicate temporary pauses in the gradual drainage of the cavities.

The events of Pleistocene (Ice Age) times formed the rich bone deposit in this cave. Figure 36 shows that there were four periods of major glaciation, while the interglacial periods were marked by quite different warm climatic conditions.

It was during one of these warm phases that the river Dart lowered its valley from the hilltop to its present height, draining the underground cave system. The uppermost part of Joint Mitnor lay only a few feet below the surface of the hill and in the next cold phase the freezing and thawing action loosened the roof blocks, which then fell into the interior. This formed an open trap for animals wandering over the hill top, and as they fell in they were buried in the earth and stones accumulating below the opening. There was, of course, no way out then, as the present entrance to the cave was only revealed by quarrying. Figure 36 shows how, in the final cold phase, the roof of the cave was sealed over again, becoming choked and jammed with boulders and stones slumping into it. So it preserved its secrets within the hill and only the deep freezing action of frost continued to affect it. To the right of the bone deposits, fragments of floor heaved up by the frost have now been recemented to each other. Like the Head of the South Devon coast and the shattered surface of Dartmoor, these blocks are another small piece of evidence as to the very cold conditions in Devon towards the end of the Ice Age. Even at this depth inside the hill the stalagmite floor did not escape.

The bone deposit forms a cone-shaped mass. Covered by stalagmite, it slopes up to the original roof entrance where roots can now be seen. The face of the cone has been excavated and prepared to show a section of the deposit. With the sterile water laid material forming the base, seven layers of deposit have been recognised and in them can be seen bones of many animals, including the vertebrae of a bison and an elephant's tooth. Other remains include hippopotamus, narrow-nosed rhinoceros, lion, hyaena, wolf, and red, giant and fallow deer. A short passage down one side of the deposit marks the site of the pre-war excavations.

Set up in this way the cave demonstrates exactly how the bones were found—it breaks away from the old idea of total excavation with an empty cavity left and everything of interest only to be seen in museum showcases. It also means that if techniques are improved in coming centuries, then another foot or so can be taken off the deposit, further discoveries about it made, and yet it will still be a demonstration site. Joint Mitnor is a small cave but it contains the richest remains of fauna from the last Interglacial (Eemian) phase yet found in Britain.

GEOPHYSICS AND CAVES

Joint Mitnor may be very much bigger than it now appears to be and the whole group of caves may prove to be a single series temporarily separated by the blocking of some passages. Surveys made in the hilltop field above the quarry suggest that there are still undiscovered air cavities within the limestones.

Individual rock formations show a differing electrical resistance, which can be measured by passing a current through the earth. The resistance recorded by caves, basically air cavities, is very high compared to the readings shown by the surrounding damp limestone rock. By working to and fro across the ground it is possible to predict where caves may be found and to detect cavities already known. At Buckfastleigh (Figure 35) further series have been predicted west of Joint Mitnor and Rift Caves and excavations in the existing caves might give access to them.

The occurrence of another cavity in the hill behind Joint Mitnor could mean that the bone deposit does not, after all, fill the back wall of the present cave (Figure 35) but occupies the centre of a once continuous system. The future development of the Cave Centre certainly promises to be as geologically exciting as its history.

Chapter 6

Paignton and Brixham

At one time the South Devon coast between the Teign and the Dart estuaries was much more regular in outline than it is today. It would have passed straight across Tor Bay from Berry Head to Hope's Nose, broken only by the mouth of the Torre Valley, and north of Torquay by the small streams flowing down between Babbacombe and Teignmouth. Today the coast alternates between bay and headland simply because of the difference between its Devonian and New Red Sandstone rocks. The softer sandstones form Tor and Babbacombe Bays while the Devonian beds, resistant to the sea's attack, stand out as headlands at Torquay and Berry Head.

Within Tor Bay itself there were many smaller-scale bays and headlands, especially in the Paignton area. Some have been obscured by the building of the sea-front roads, but many were already disappearing naturally long before that. Salt marshes accumulating in the shallow bays and inlets gradually filled them in. Extending seawards, they reduced the prominence of intervening headlands like Roundham and Livermead. The results of this activity can still be traced today.

In Paignton, if you stand on any kind of slope, it will be formed on the New Red Sandstones, but if you are on low-lying flats like the seaward end of the town centre then the ground was once marshland.

The historical evidence is very convincing. The old village of Paignton stood on firm ground on the inner margin of the marsh—the Church and Kirkham House area. The roads leading to it also kept to dry ground, eg, the old Torquay road through Preston. Although the marshlands were reclaimed in the last century and the town extended seawards, the water level still lies close to the surface even as far inland as the bus-station.

Much of Paignton's geological interest lies between Roundham Head and Broadsands. In the cove east of the harbour a junction of New Red Sandstone formations can be seen. The base of the Roundham Head cliff here is in Tor Bay Breccias. Dipping gently northwards from Waterside Cove, they form the whole southern cliff of Roundham and finally reach sea level here, creating a bouldery wave platform. The bulk of the cliff above is in the next higher formation, the sandy Livermead Beds. They are not very resistant beds and form low-lying areas in other parts of Paignton, in the town centre and behind Preston Sands. Here, in the eastern face of the cove, cross bedding can be seen in the Livermead Beds—they appear to have been laid down in groups of curving structures. The feature is due to slight changes in the direction from which the material was arriving (Figure 37).

Higher up, the cliff-face looks rather like a brick wall with

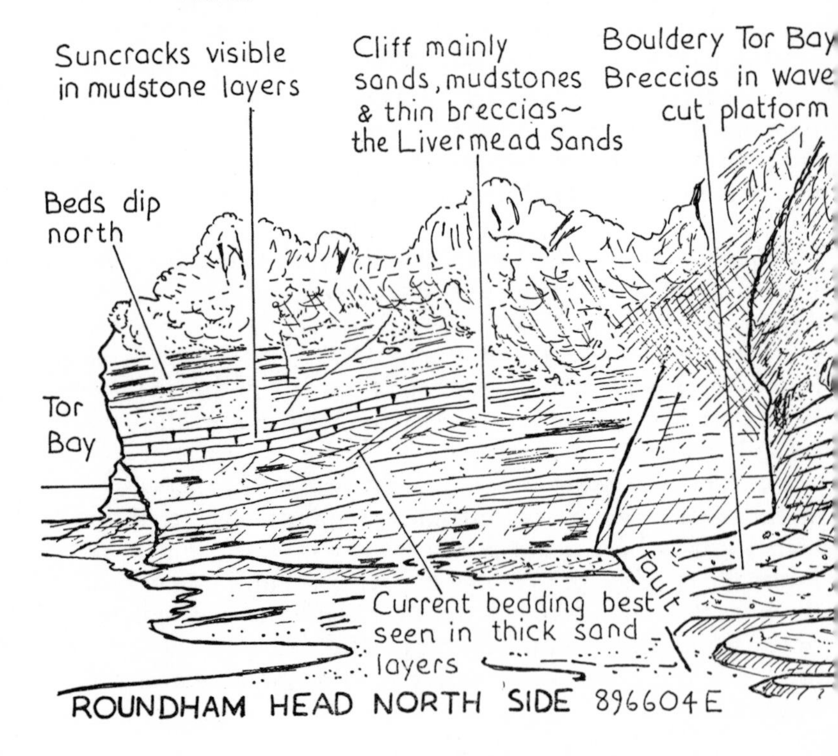

regular horizontal and vertical ribs standing out in muddy beds. When the beds were on the arid surface they dried and cracked in the sun, their surface looking like crazy-paving. Sand, washed over the surface later on, was able to fill these tapering cracks, so creating the ribbed appearance seen in the cliffs today. Since the mud-cracks would naturally be wide at the top, these ribs can be useful evidence that the beds are still the right way up and have not been overturned by later earth movements. In the middle of this cove beekite-covered pebbles can sometimes be found (Page 66).

GOODRINGTON AND WATERSIDE

Walking over the headland to Goodrington Cliff Gardens there is a good view to the south. The gardens stand on the rough Tor Bay Breccias, although at the end of the promenade some sandy-orange beds can be seen. The Goodrington beach cafés and hotel were built on a small sandstone outcrop once surrounded by the flat land of salt marshes extending half a mile inland along the Clennon Valley. At the southern end of Goodrington Devonian Beds outcrop again, much folded and faulted with small areas of New Red thrown down along the coast.

There are many coastal paths between Goodrington and Broadsands, so access to short sections is possible at all states of the tide. At low tide the shore can be followed all the way to Broadsands.

It is easy to confuse the names of the coves along this coast. Waterside is the middle one of three to the south of Goodrington Sands. Starting from Three Beaches, go up past the church and cross eastwards over the railway and the field beyond, but if the tide is out try the walk along the shore from Goodrington. A puzzling feature is encountered in the blocks of sandstones and breccias lying at the base of the cliffs. The pebbles in them seem to have been rearranged by a burrowing creature (Figure 38), the only fossil evidence found in the local New Red beds. If these organisms were worm-like creatures, then they must have been at least six inches thick and could apparently eat small pebbles! They liked finer sand and coarser pebble beds equally it seems. Another more likely explanation could be a burrowing animal

FIG 38

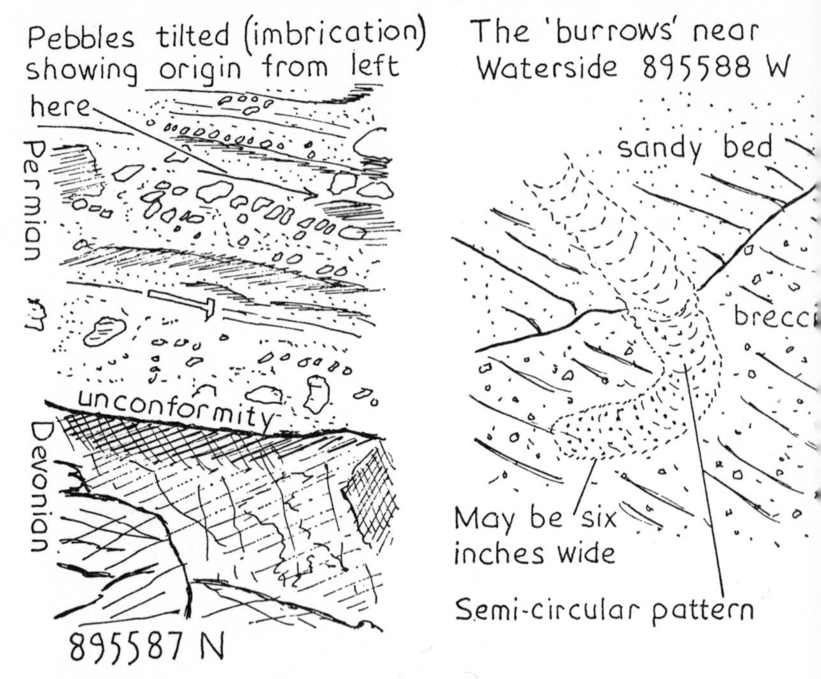

FEATURES NEAR WATERSIDE COVE

or reptile, but there is no way of identifying it and the type of life which produced these features remains a puzzle.

In Waterside Cove itself there are Devonian purple slates and sandstones, cut across at the northern end by a remarkable unconformity where coarse limestone breccias of the New Red Sandstones lie on an eroded surface. Walk into the undercut of the cliff (Figure 38) and put your fingers on the under-surface of the breccias. A negative impression of the old slate land surface can be seen. You will have your hand on a surface that represents the time interval of the whole of the Carboniferous period, all the beds of which were removed by erosion before these New Red Sandstone breccias were laid on top.

The pebbles in the breccias above the unconformity are interesting. Close inspection shows how they are slightly tilted up against each other. Figure 38 shows how this effect reveals the direction they came from. The feature is termed 'imbrication' and can be seen in many gravelly stream deposits today. Crossing the beach, it is worth spending a while on the seaward view. In sunny weather the contrast of red cliff and the blue waters of the bay is vivid while beyond are the enclosing arms of white limestone at Torquay and Berry Head.

On the south side of the cove faulting brings the explorer back into Upper Devonian rocks again. The rocks are red and muddy, with a strong horizontal cleavage which almost hides the fact that the bedding (revealed by whitish bands) is nearly

MAP 8

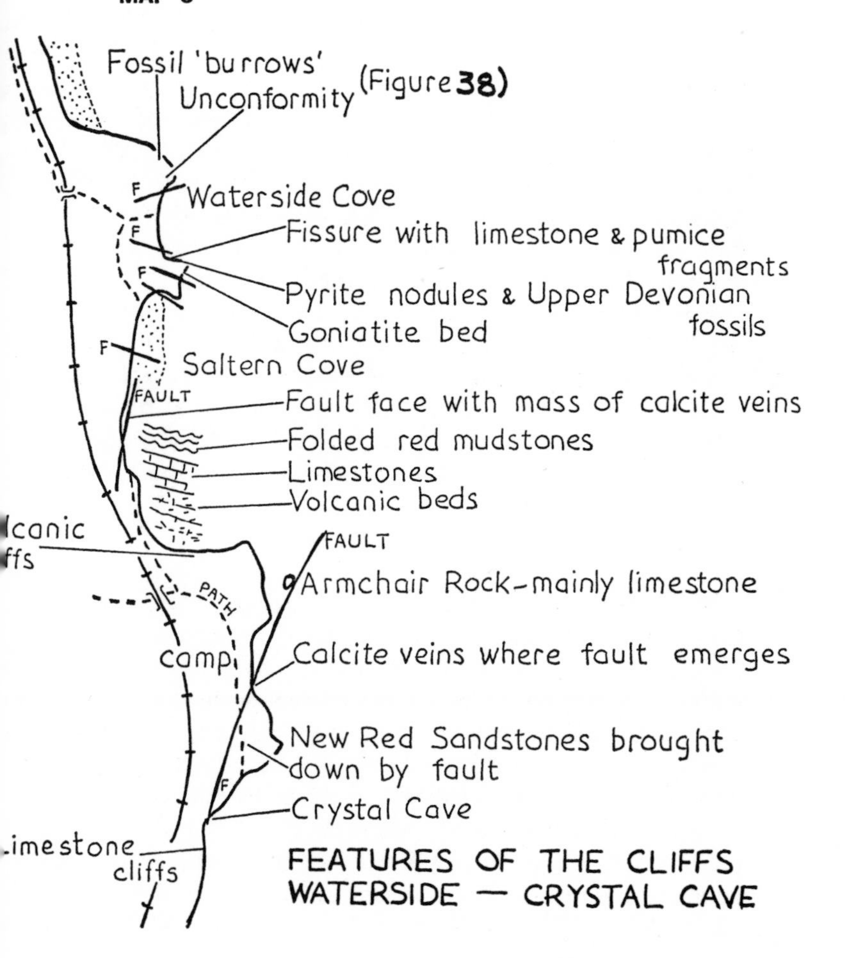

FEATURES OF THE CLIFFS
WATERSIDE — CRYSTAL CAVE

vertical. Running up the cliff and projecting onto the beach below is a 2-foot-wide dyke with fragments of pumice and limestone in it. It was some sort of volcanic outlet.

The red mudstones contain many black nodules of pyrite, and after looking at these for some time the explorer will notice that some of them are really very small fossils. Tiny goniatites, trilobites and lamellibranchs weather out here, a typical Upper Devonian fauna noted all over north-west Europe for its stunted size. They belong to what geologists term the *Manticoceras* zone after its type fossil. One of the most common creatures here is the little lamellibranch *Buchiola retrostriata*—but samples barely reach one-eighth of an inch across. Goniatites can also be found on the tip of the headland between Waterside and Saltern Cove, the next bay to the south.

Faults can easily be traced in Saltern Cove where they have affected the Devonian limestones and volcanic beds. One forms the cliff face and, beginning at a limestone outcrop in the centre of the cove, it can be traced southwards as a reddish mass of calcite veins for well over a hundred yards. The beach is interesting here, too. Along the shore near the fault a sequence of mudstones, limestones and volcanic rocks can be traced southwards (Map 8). Volcanic cliffs border the south side of the cove while the limestones appear again in the isolated stack of Armchair Rock. Climbing up to the caravan camp, the walker should keep to seaward of the railway and follow the path down to the northern end of Broadsands.

THE CRYSTAL CAVE

Of all the small faults which exist along this part of the shore, the most famous must be the one responsible for the Crystal Cave. Crossing the lower field of Waterside Camp, it emerges in the cliff face in a smashed zone nearly 30 feet wide. The whole area around it is thickly webbed with calcite veins, the millions of crystals distinctively shaped by their three natural cleavages. Within the cave every wall glitters with shiny facets, although a great deal of damage has been done by the sea and by collectors. This is a place for the camera rather than the hammer.

FIG 39

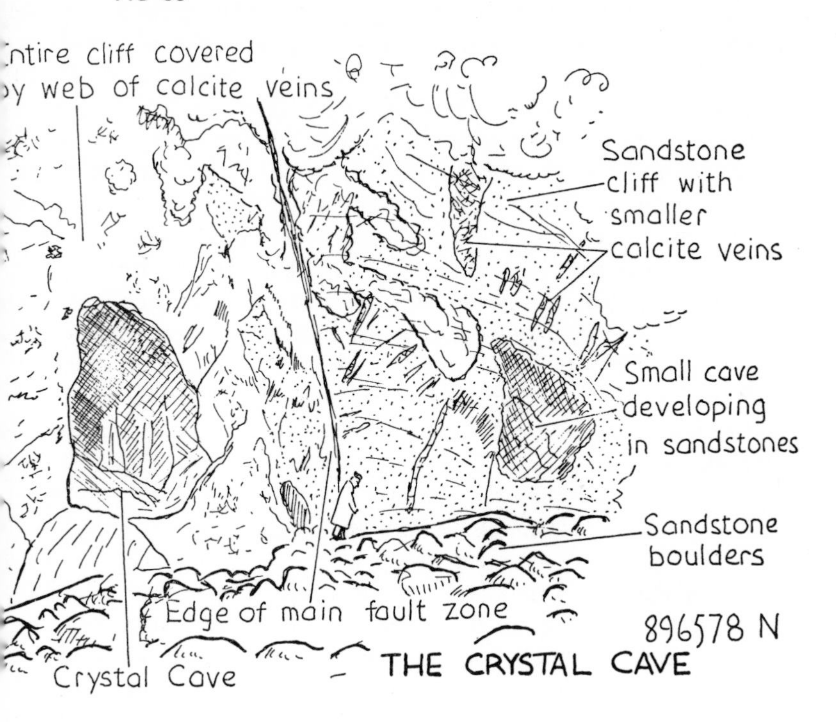

BROADSANDS AND THE DART

At Broadsands, the explorer reaches the great limestone plateau which marks the southern shore of Tor Bay. Its flat 200-foot marine surface extends from Berry Head to Churston and Galmpton.

In times when its valley was at the same height as the limestone plateau, some geologists believe the river Dart flowed out towards Brixham. Other streams crossing Tor Bay would have been its tributaries. But this argument has a serious weakness—no one has yet found river gravels in the Churston or Brixham area which would provide proof. Is it coincidence that an old level of the Dart valley and the limestone plateau are at the same height? And if the Dart did go to Brixham, why did it make a change later? One possible explanation is river capture. Another stream cutting the deep gorge at Dartmouth could have worked into the Dart valley at Galmpton and diverted it. Those who believe there is no link between the Dart and the plateau regard the latter as a purely marine-cut feature like the levels around Torquay. At 200 feet above sea level it does agree with the heights of Daddy Hole Plain and Waldon Hill.

The tower of an old windmill on Churston Common illustrated a common problem of limestone districts which stand above the water-table, their lack of surface streams. There was no stream in Churston parish to sustain a watermill.

The only sizeable valleys in the plateau are the ones which meet in Brixham and provide the town with its sheltered site. Many quarries have been worked on their slopes near the town and on the seaward margins of the plateau around the harbour. The working faces emphasise the erosional origin of the plateau above, which clearly cuts regardless across all the rock structures.

BRIXHAM AND IRON OCHRE

On the way into Brixham the site of the former Tor Bay paintworks stands on the right (now Decca Navigation). The original works was near the harbour, perched directly above the sea by Freshwater quarry. In 1842 a local man discovered rich iron-oxide deposits in the limestone plateaus, and so began the paintworks which became the Tor Bay Paint Co in 1895, by which time it was exporting its product all over the world. The paint was valued for its preservative qualities. Discouraging rust, it was in demand for many metal bridges, and among the contracts held by the company was the maintenance of Victoria Falls bridge on the Zambesi river.

The basic treatment was first to dry the ochre in a pan kiln similar to those in the old china-clay dries and then to roll and crush it to a fine powder before mixing it with varnishes and thinners.

The ochre was obtained from opencast works on the side of Rea Hill. Several hundred square yards of workings were dug by the north side of the St Mary's Bay (Mudstone Sands) to Higher Brixham road. The deposits were shallow, 20 feet thick at the most. More ochre pits existed above the harbour—about midway between the original works and the now vanished railway line.

MAP 9

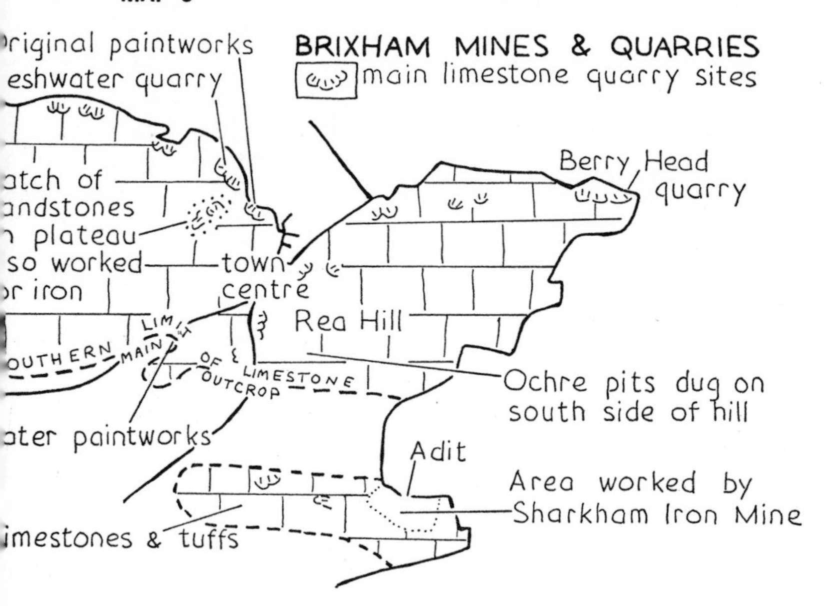

The paintworks had a second source of iron in the red and brown haematite veins worked near Sharkham Point to the south of the town. There iron solutions had replaced parts of the limestone and created irregular earthy deposits. Only the soft material went to the paintworks, the harder going for smelting in South Wales. Overgrown remains of mine shafts can be traced around the north side of Sharkham Point where one adit cuts through the limestone bench of a raised beach. Standing 15 feet above high water, the beach is now partly obscured by mine waste dumped on top of it.

SHARKHAM, BERRY HEAD AND SHOALSTONE

Formed of limestones, Sharkham Point provides views of the coast to north and south. The thin-bedded limestones can be traced across its flat summit, running parallel to the axis of the headland. A mass of iron-impregnated volcanic rock associated with Sharkham iron mine can be seen on the north-east cliff face.

The steeply inclined, pinkish slates and grits to the south are reminiscent of Blackpool Sands (Page 31) and four small sea-caves in the slates at Crabrock Point can be visited at low tide. The beds here are Lower Devonian but north of Sharkham in St Mary's Bay there are grey Middle Devonian slates. St Mary's Bay has been cut back in these weak beds, between the limestone arms of Sharkham and the Berry Head area.

Beyond St Mary's, the coastal platform is very fine and the way it cuts across all structures is obvious. Here the most interesting group of sea-caves in the area can be found. Climbing down the grass-covered slopes 300 yards west of

FIG 40

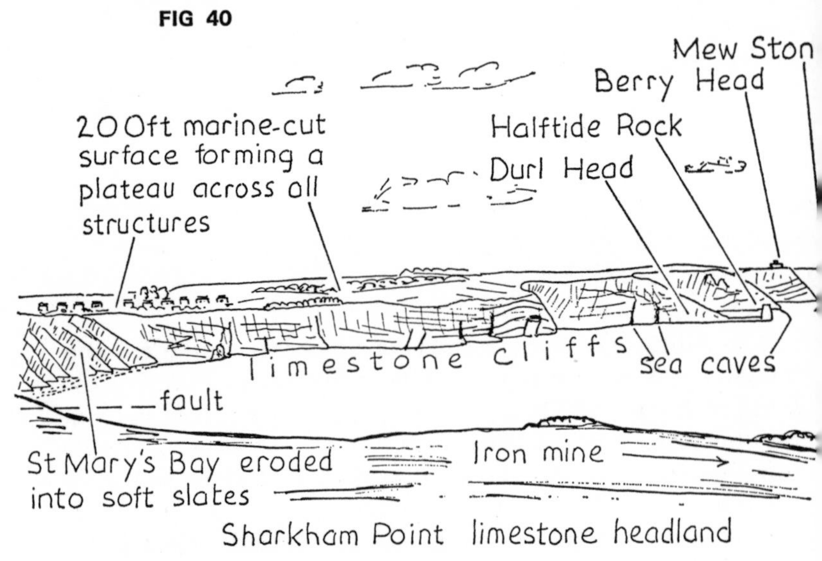

ST MARY'S BAY – BERRY HEAD 935547 N

Durl Head, there is a cave with two entrances. One entrance is partly blocked by cliff falls but the other is open to the sea. A considerable beach lies inside but the main interest here is the rapid stalagmite growth. It forms right down to sea level, coating glazed pottery fragments which have fallen into the cave and even mosses growing near the entrance.

Another unusual cave occurs in the point opposite the Mew Stone. With two entrances on the north side and another on the south, a rowing boat can be taken through it at low tide.

On the south side of Berry Head the natural cliffs have an unbroken rise to the marine platform 187 feet above, but on the northern side extensive quarries have destroyed their original form. Raised beaches cannot be traced right around the head but they appear again at Shoalstone beach. This beach is a 'must' on the walk back to Brixham because there is also a wonderful set of sandstone dykes there.

The contrast in colour between the red dykes and white limestones makes the locality a very good one for mapping exercises. The dykes accumulated in two stages when the whole area was being stretched by earth movements which opened up fissures in the limestones. At first, the fissures must have only contained groundwater for they became coated with a creamy lining of crystalline calcite. Then the first red sands were washed in—derived from local New Red Sandstone beds. They were also cemented together by calcite in the fissures.

FIG 41

THE SHOALSTONE BEACH DYKES 936568 E

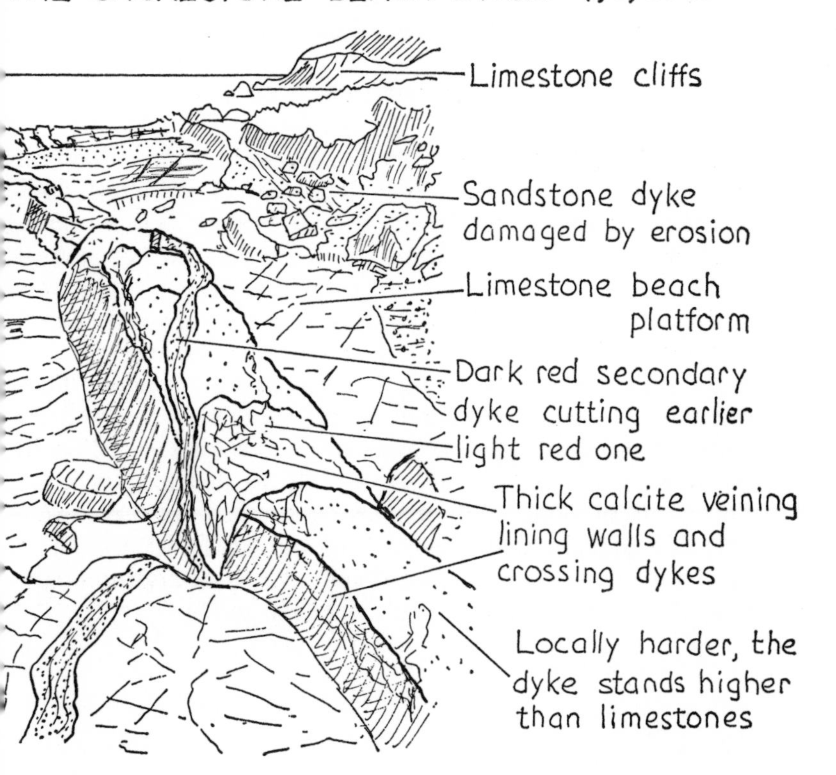

The sands must have solidified very quickly because when renewed movements opened the fissures for a second time the pale red dykes remained in position. If they had still been loose they would have slumped down to the bottom in each case. Instead they split open themselves to receive another coating of calcite followed by darker red sands. Both light and dark groups can be distinguished on the beach, and the paths of younger dykes traced through and sometimes across the older ones.

The dykes generally lie N-S or E-W but they do not fill the entire length of each fissure. Obviously they were only washed in where opportunity allowed. The dykes are very extensive vertically though—in nearby Berry Head quarry and others in the district they can be traced right up to the plateau level above. So the sands in Shoalstone beach must have descended through the entire thickness of limestone present in those days. A small quarry, now an extension of the Shoalstone beach car-park on the upper side of the road, provides a convenient example of a vertical dyke exposure.

BRIXHAM CAVERN

The cave entrance lies beneath a house in Mount Pleasant road. The road was cut along the slopes of Windmill Hill in 1840 and nearby land worked for limestone before houses were built on the site. Many Brixham houses are perched on the rock shelves of old quarries in the hillside, and it was during construction here in 1858 that the cavern was found.

Lying 100 feet above sea level, the cave was excavated under the supervision of the Geological Society of London. Four levels of deposit were found. Man's flint implements lying with the remains of mammoth, rhinoceros and cave lion in the third bed gave conclusive proof of the antiquity of man. Fragments of dolerite in the lowest beds in the cave were probably brought there from the Sharkham Point area. The cave is well worth visiting for despite the necessary thoroughness of its excavation, there is still much to see.

Chapter 7

Teignmouth and Dawlish

Oddly enough, 'Red' Devon has much more grey and brown coloured rock than red. But the description doubtless comes from the vivid impression travellers obtain along the roads near Exeter, or from the railway line between Dawlish Warren and Teignmouth. The cliffs of brightly coloured sands make a startling contrast with the rich greens of the fields and the blue sea. Their outcrop follows the Exe Valley inland, running into Somerset and extending in finger-like outcrops westwards from the river to Hatherleigh and Tiverton. But like so much Devonshire geology, the best exposures are along the coast where there is an almost unbroken section from Torquay to Branscombe. Other outcrops, further south, are described at Paignton and Slapton.

The red sandstone coast has several remarkable features. Its direction is fortunate—cutting across the old semi-arid basin in which the rocks accumulated, it reveals the changing nature of the beds as the explorer moves in from the margins to the central area. The rivers Teign and Exe provide interesting spit formations in the Den and Dawlish Warren, while south of Shaldon the cliffs illustrate the balance between the sea and other types of erosion and the development of hanging valleys as the coast recedes.

THE PERMO-TRIASSIC BASIN

The New Red Sandstones include rocks of both Permian and Triassic age. Locally, the division between the two has been traditionally accepted as the base of the famous Pebble Bed at Budleigh Salterton. However, recent research suggests that the boundary between the two periods would be better placed at Exmouth.

The beds accumulated in arid basins similar to those of

FIG 42

THE LANDSCAPE OF PERMIAN TIMES

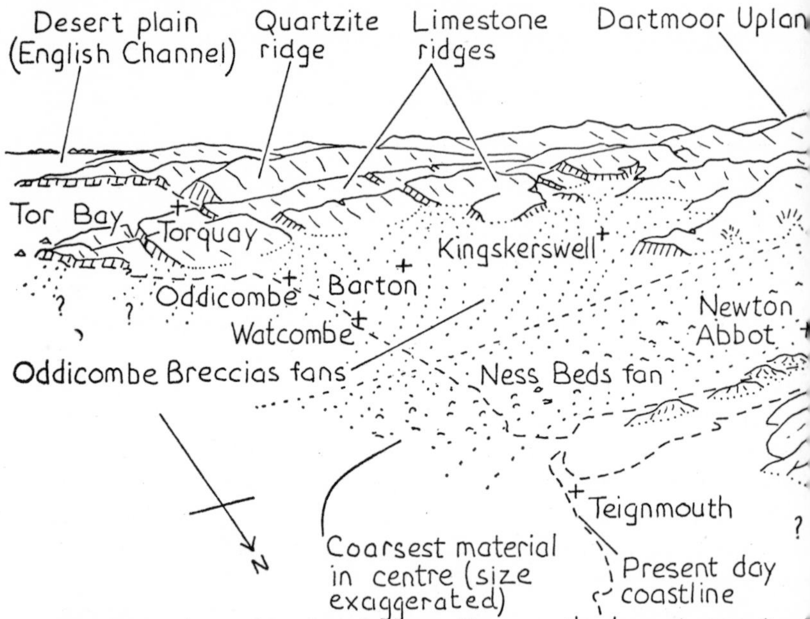

Limestone breccias forming on the south do not penetrat the central area much. Ness Beds derived from Dartmoo Upland area are porphyry, &c. A low slate ridge marks site of future Teign estuary. Present day Devonian ou crops at Barton not shown as they were lifted by later earth movements

modern Persia. Figure 42 illustrates the conditions, with the older Devonian and Carboniferous rocks forming the surrounding mountains. The climate was dry, but not as hot as the modern Sahara Desert. Occasional rainstorms flushed away rock waste from the ridges and carried it out into the plain. In dry periods, and particularly out in the centre, winds blew the loose surface material around and dunes up to 60 feet high were formed (see Coryton's Cove, Dawlish, below).

Basically, three types of New Red rocks were produced—wind-blown sands, sandstones laid down in water, marls and breccias. Breccias with a coarse mixture of other rock fragments were laid down by flash floods. The explorer can soon develop an eye to distinguish between the three types and so visualise the conditions in various parts of the basin.

Apart from this there are two other features to familiarise oneself with. Firstly there are differences in the breccias from different parts of the basin. In the central areas around Teignmouth and Dawlish the fragments arrived along the main drainage route, pieces of Carboniferous sandstones and metamorphic rocks from the roof which still covered the Dartmoor granite in these days. On the margins, however, the breccias were more local—the southern ones contain limestone and quartzite fragments from the Devonian ridges of Torquay and Newton Abbot, for example. It needed an exceptionally heavy flood to carry these pieces as far as the centre of the basin, as their less frequent discovery at Dawlish testifies.

There are also differences due simply to the gradual filling up in the basin. As the level of the debris rose, so it helped to bury the surrounding ridges and reduce its own sources of supply. The flatter form made it even more difficult for anything but the finest materials to get far out into the centre. Wind action on the material already in the basin led to sand-dune formation, and their forms can be recognised in the eastern exposures today. So generally the New Red beds become finer and finer as the walker moves northwards and eastwards into the old centre of the basin.

WATCOMBE TO SHALDON

The best thing to do is to start at the edge of the New Red basin near Watcombe. As far as Shaldon the coast path can be rather exhausting. The switchback form and constant ascents and descents soon impress on the walker the nature of a cliff notched by hanging valleys. Some sections like the Goat's Path at Maidencombe demand a head for heights, although the path is wider now than it used to be. The main road between Torquay and Shaldon keeps inland, avoiding the steep gradients by following the ridge between the coastal combes and the streams which flow to Combeinteignhead.

At the beginning of the walk the New Red Sandstone can be viewed from Watcombe beach car-park. The vertical face is spotted with numerous lumps of limestone. Its peculiar honeycombed appearance, common along the coast, is the work of wind erosion and is described in exposures at Budleigh Salterton (Page 129).

The Watcombe-Shaldon coast is a Devonshire version of the famous Seven Sisters near Eastbourne—the nearly vertical cliff is notched by a series of valleys which fail to reach sea level and are left 'hanging' in its face. The upright cliff profiles show that wind, rain and frost attacking the upper cliffs are working on them at much the same speed as the sea attacking the base.

The hanging valleys are the remnants of streams which once continued normally down to sea level. Imagine their gradients extending beyond the present cliffs—they would reach the sea about a mile farther out. The lower half of each valley has disappeared with the erosion of this coast and the

FIG 43

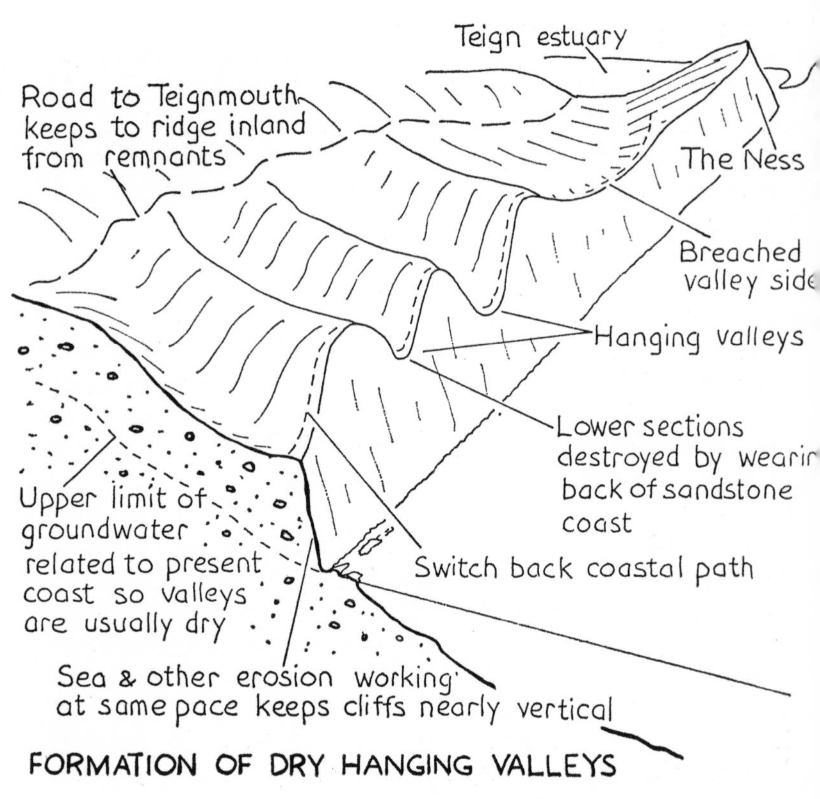

FORMATION OF DRY HANGING VALLEYS

surviving upper areas have lost their streams as ground water levels adapted themselves to the new shoreline (Figure 43).

The Ness at Shaldon is an unusual shaped headland which has some romantic appeal in its Smugglers' Tunnel. The tunnel served to bring limestone up from the beach to the kiln at its inland entrance—at least by day! The shape of the Ness is another variation on the theme of coastal erosion and hanging valleys. Here a small stream, rising about three fields to the south by the main road, flowed north to the Teign. Its valley therefore ran parallel to the receding coast instead of towards it, and so, in this case it lost part of its side. Hence the shape of the Ness. The seaward slope is the newly eroded cliff face, the inland one a surviving piece of the valley side (Figure 43).

On the Ness beach below the tunnel the breccias include fewer fragments from the margin of the basin, and many of those seen here are surrounded by greenish patches due to the presence of ferrous iron minerals. The normal ferric iron in the sandstones was changed by the lime content of the pebbles.

There is much more breccia material now from the basin's central sources. Quartz porphyries, often reddish brown with glassy quartz crystals, and pink or white felspar patches are common. They came from the now-eroded roof of Dartmoor some miles distant to the west. Violent floods must have occurred to bring them here. One fallen porphyry boulder on Ness beach was five feet across and larger ones have been found inland.

THE TEIGN AND THE DEN

Teignmouth is a busy port for its size, handling the ball clay trade from the Bovey Basin. The clay used to come down river by barge but the channel along the river was difficult to maintain even in the last century. Like most South Devon rivers, the Teign lacks sufficient power and a good scouring tide to remove the sediment once it has formed in quantity. The view from Bishopsteignton towards Newton Abbot confirms the continued rapid infilling of the estuary. At Teignmouth, barges and sea-going ships tied up alongside each other in the river to tranship the clay cargoes for quay facilities in the port were poor before 1886. Even with improvements made since then, the harbour remains a difficult one for shipping to enter. The Den, the curving channel on its inner side and the bar at the mouth are all products of the relationship between the river, wave action and the amount of sediment in the estuary. On this part of the coast the strongest waves reaching the shore are those built up and accompanied by easterly winds. Recently the easterly winds have become much more frequent, generating waves which have done a great deal of damage to Oddicombe beach, Dawlish Warren and the Den.

The Den is now quite built up, whereas originally it was reed-covered marshland, a part of the manorial waste used for drying nets and as a promenade. Early settlement was well inland at the foot of drier ground. The Den was gradually enclosed for building from the early nineteenth century but despite its man-made additions the sea is still quite

FIG 44

NEW RED SANDSTONES SOUTH OF CORYTON'S COVE 962760S

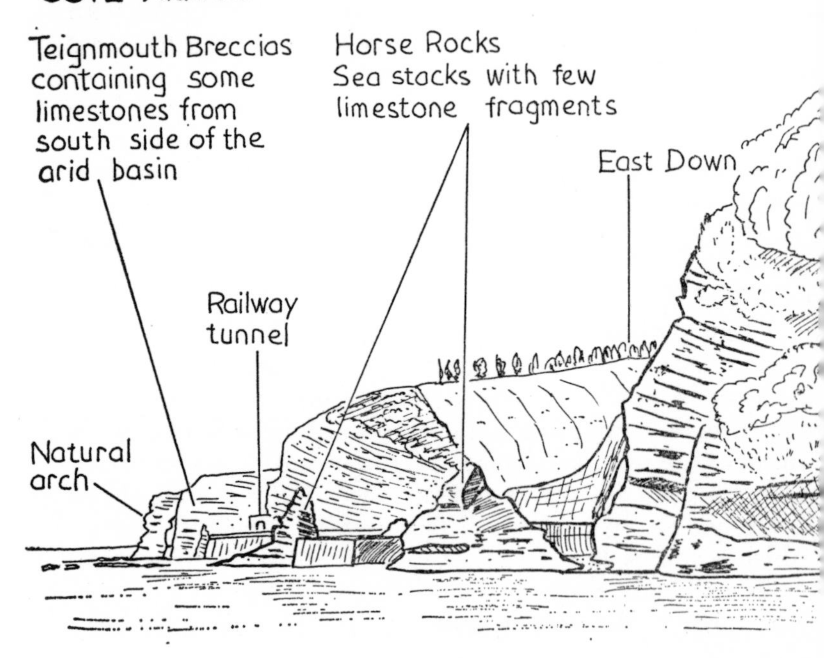

capable of reclaiming it. When wind and tide are right, severe flooding occurs over it and into the low-lying town centre.

CORYTON'S COVE, DAWLISH

By the time the walker reaches Dawlish he is so well out in the centre of the old basin that wind-blown sands become a feature of the New Red beds. At the south end of Dawlish

FIG 45

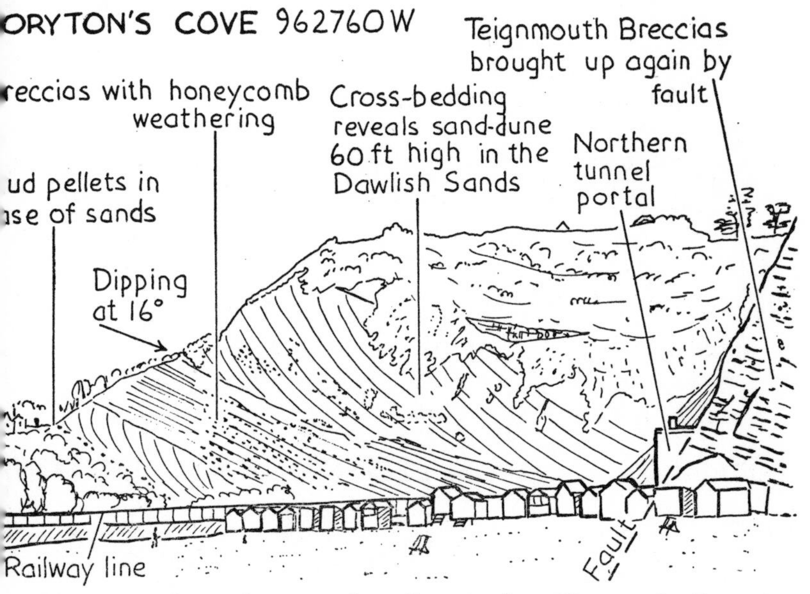

front, cross the railway and walk round to Coryton's Cove. More honeycombed cliffs can be seen on the way.

The Teignmouth Breccias dip below sea level at Horse Rocks just south of Coryton's Cove and the cove itself lies in the next higher formation, the Dawlish Sands. These reveal the fossilised form of probably one of the largest sand dunes which existed in those days. Formed by wind currents which built it, the cross-bedding marks in the cliffs above the railway line reveal that it reached a height of at least 60 feet. Local measurements of cross-bedding reveal that the wind in those days blew from the south-east, quite different from today's prevailing south-westerly wind.

FIG 46

REPETITION OF BEDS DUE TO FAULTING

An undisturbed sequence

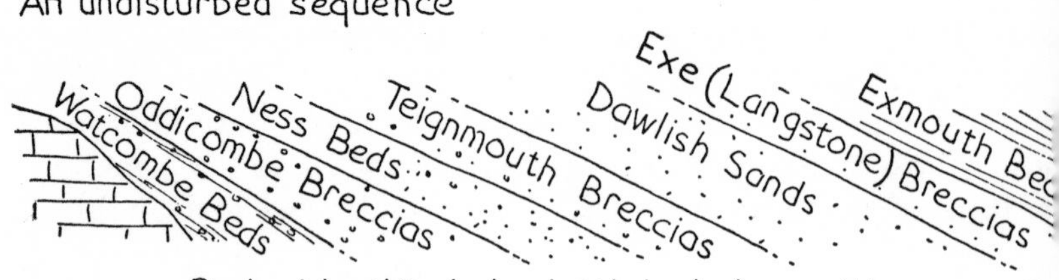

Beds identified by initials below with generalis
outline of the coast north of Torquay

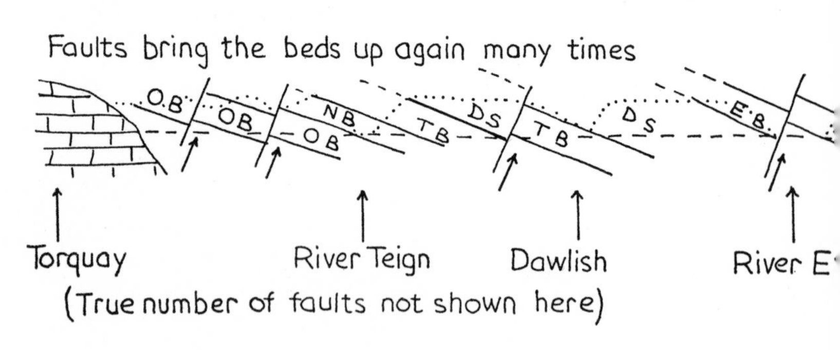

FIG 47

FORMATION OF CROSS-BEDDING BY WIND EROSION

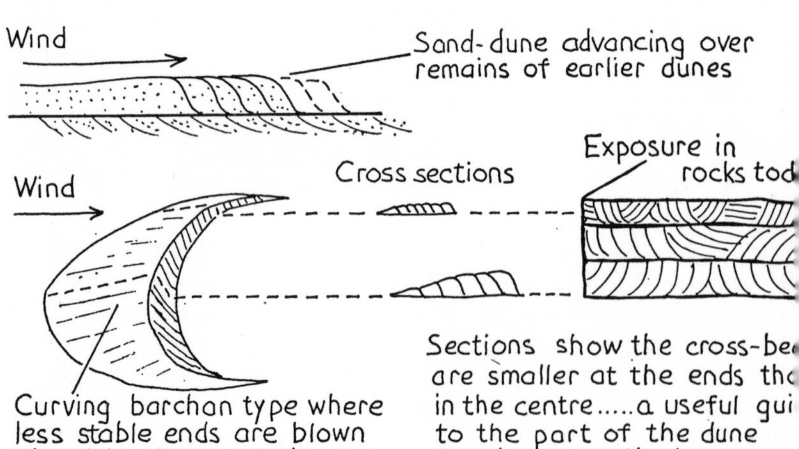

Sections show the cross-be
are smaller at the ends th
in the centre.....a useful gui
to the part of the dune
forming a particular exposu

The exposure of the sands ends abruptly at the portal of the northern railway tunnel. There the Teignmouth Breccias appear again, brought up by a large fault with a movement of nearly 200 feet. The fault lies exactly on the position of the tunnel entrance and, much-altered by erosion, its face also forms the northern side of the cove. The presence of so many faults of this type explains why various New Red beds keep recurring in the cliffs, long after they would have otherwise dipped below sea level.

Dawlish Water and the town centre probably lie on one such fault and there are others to be spotted on the coast path to Langstone Rock.

FIG 48

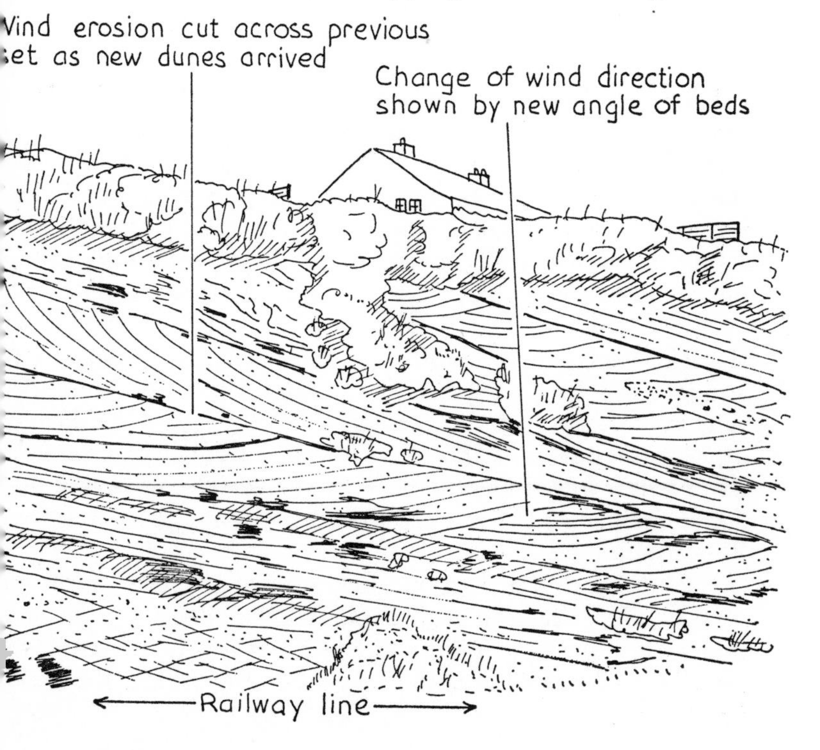

It is worth stopping to look at the cliffs by the footbridge below Rockstone Hotel. Just south of the bridge there are several small faults and the cross-bedding shows small triangular patterns. These represent the tips of the original sand dunes whereas north of the bridge larger, curved cross-beds indicate the central areas of dunes (Figure 47). More faults and gullies occur near the platelayers' hut and beyond this point the character of the Dawlish Sands begins to change. Water-laid beds appear more frequently, the work of streams attacking and re-sorting the dunes.

An actual stream channel is exposed about 50 yards north of the hut. Filled with coarser, pebbly material, it begins as a trail of stones arriving from the south, thickening into the channel almost at track level.

Nearer Langstone Rock the cliffs are capped with a yellowish gravel deposit which relates to an old estuary level of the river Exe. It also occurs on Langstone Rock itself and can be traced along the hillsides towards Powderham.

LANGSTONE ROCK

Langstone Rock is a good vantage point for Dawlish Warren. To the south, the smooth coastline of the railway embankment gives way to headlands beyond Dawlish, with the famous stacks of the Parson and the Clerk. Like all the formations of this sandstone coast these two features have subtly but steadily changed over the years and some interesting hours can be spent comparing early picture postcards of them.

Brightly coloured and flat-topped, Langstone Rock has a slab-like appearance. It reveals a blow-hole, small sea-caves and joints to those who take time to explore it. Long before the railway severed it from the cliff-line, it was under heavy attack from the sea. Parts of it are still, but much of the Rock has been involved in attempts to protect the railway line. These measures have affected Dawlish Warren as well as the Rock itself.

Exe Breccias, outcropping in its southern face, form a vertical cliff where landslips occur at intervals as the sea undermines the base along some small faults. Faults are also responsible for the formation of a natural arch near the little breakwater.

FIG 49

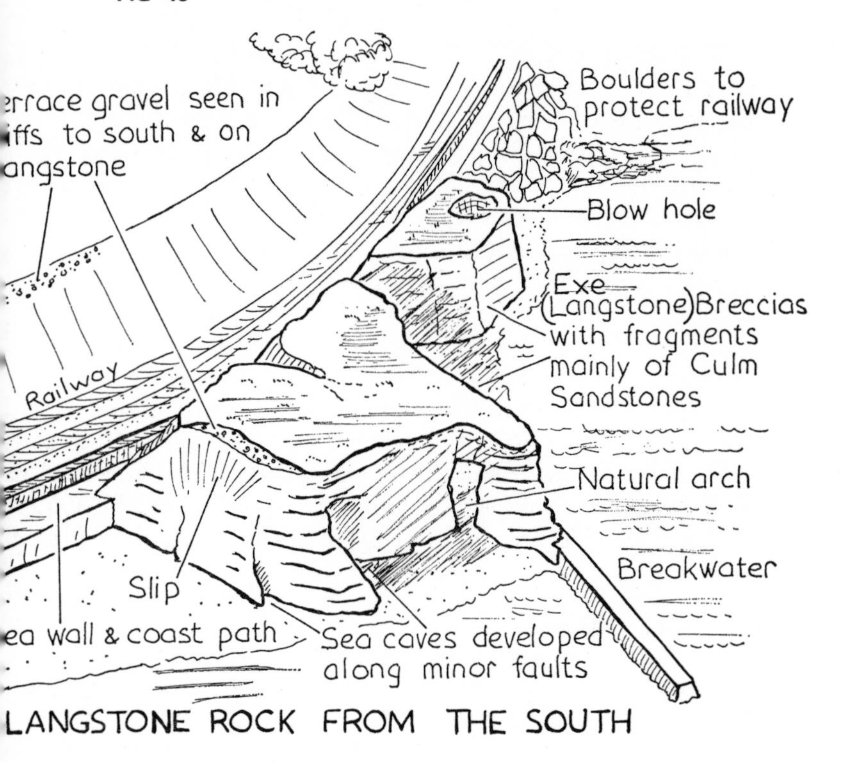

LANGSTONE ROCK FROM THE SOUTH

Before the railway was built, erosion of Langstone Rock and the coast to the south provided a good deal of sand for Dawlish Warren. The railway cut out most of this supply and the small breakwater, designed to protect the railway, aggravated the position still further. The little material there was could not pass north-east around Langstone Rock. The breakwater also concentrated easterly wave attack against the north face of the rock, creating a large blow-hole there and making additional protection necessary for the railway embankment. This was provided by tipping large granite blocks.

The railway has certainly aggravated the erosion of Dawlish Warren, but in fairness it cannot take all the blame for records show that the Warren was also heavily eroded in the sixteenth century.

DAWLISH WARREN

Flat and sandy, with bracing winds, golf course and camp sites nearby, Dawlish Warren has many attractions for the tourist. But it is a very unsafe bathing area, for the tip of the Warren is affected by strong tidal currents entering and leaving the Exe estuary. Various systems of groynes, designed to prevent the waves stripping the Warren of its sands, have not been very successful.

For much of its recorded history the Warren has been a double feature—an Outer Warren of broken sandhills separated by a creek known as Greenland Lake from the Inner Warren, grass and gorse-covered and the site of the golfcourse. The Outer Warren has been repeatedly reduced in area since the end of the eighteenth century, while the end of the spit has advanced and retreated several times in the direction of Exmouth. For a time it seemed that erosion of the Outer Warren was helping to build up Warren Point, the extreme

FIG 50
DAWLISH WARREN

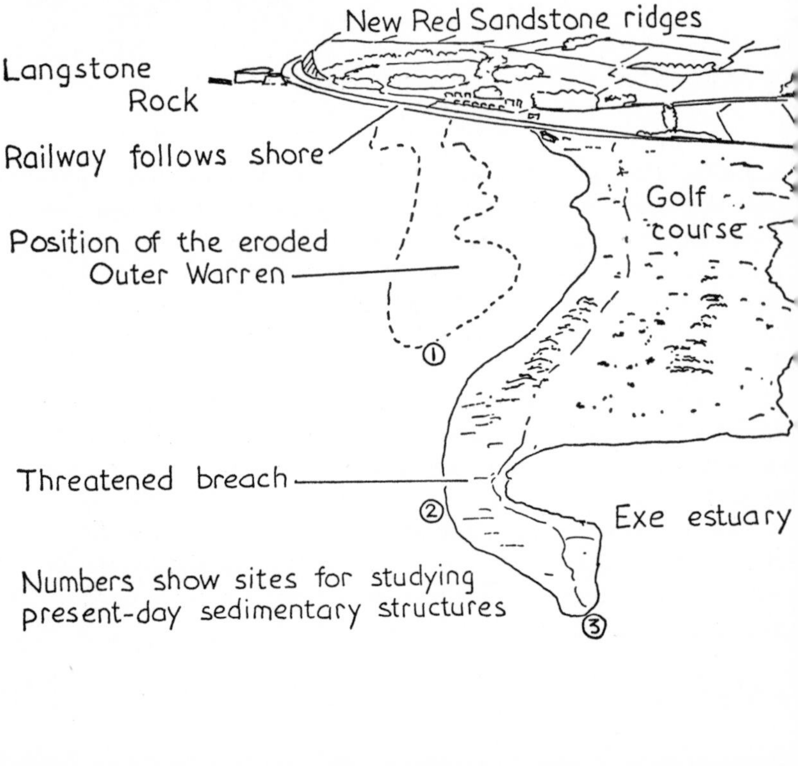

New Red Sandstone ridges
Langstone Rock
Railway follows shore
Golf course
Position of the eroded Outer Warren
1
Threatened breach
2
Exe estuary
Numbers show sites for studying present-day sedimentary structures
3

end of the spit. In 1938 the channel across the river Exe was reduced to the lowest width ever recorded, 253 yards. But even at that time Warren Point was being attacked from seawards and a colony of bungalows there gradually disappeared between 1923 and 1946. It was in 1946 that Warren Point finally broke up and the channel widened to over 600 yards. Loss of the Outer Warren has now brought the Inner Warren under attack and largely obliterated Greenland Lake.

The remains of a great variety of groynes can be seen on the seaward shore, and the continuing erosion illustrates the problems of designing adequate groynes. Figure 4 on page 21 shows the method by which beach material migrates along the coasts. The simplest wall-like groyne is merely a barrier to this movement, the sand piling up against the obstacle. More complicated ones have been tried on the Warren. With open bars they were designed to slow the speed of attacking waves, preventing them from scouring the sea bed and making them drop any sand they were carrying. Remains of wattle fences aimed at controlling wind-blown movements can also be seen.

It is difficult to assess where the problems of the Warren really begin—at its tip, on the seaward shore, or at Langstone Rock? The ebb-tide flow, sweeping in violent eddies around the end of the Warren, sets down its seaward side in strongly scoured channels parallel to the beach (Figure 51). These currents, so dangerous to bathers, reverse the sources of supply and lately the changing outline of the spit has been so badly damaged that it has almost been breached again near the outer end. Plans are now being implemented for new major groynes near the outer end. These will slow the ebb flow but the late decision on the site has been serious and meanwhile erosion goes on rapidly.

Within the Exe estuary the currents carry out a good deal of natural sorting on banks like Bull Hill. The flood-tide sweeping in across Bull Hill at its highest moves gravels directly over it, the ebb-tide then reworking the finer sizes back round the bank in an anti-clockwise direction. Mussel beds on these banks have varying degrees of success in stabilising them, sometimes losing their foundations by undercutting of the sands and at others being killed by new banks advancing over the top of them.

FIG 51

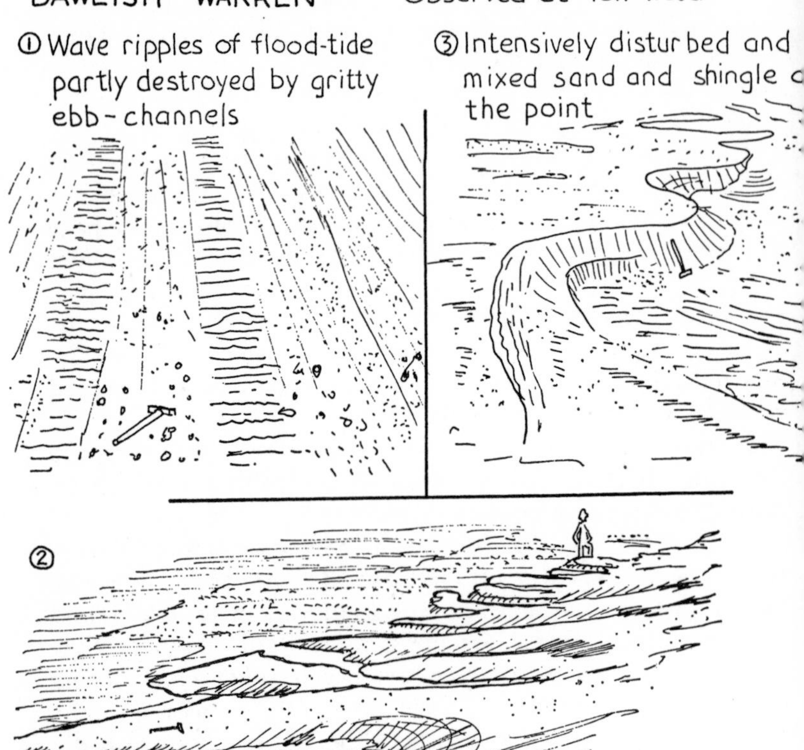

Geophysical studies of the banks suggest that in the New Red Sandstones below, the Exe has cut many channels. The deepest set would be related to the minus 150-foot level of the sea described in Chapter 2. Their number may be due to several zones of weakness associated with the Exe estuary fault described later.

Chapter 8

Exeter and Its River

A car journey across the Exeter district from Bickleigh to Exminster soon reveals its basic geology. High ridges of Carboniferous (Culm) shales and grits alternate with outcrops of softer New Red Sandstones. The two formations are intertwined across this part of the Exe Valley.

An instructive view can be obtained from Yards Downs above Silverton. At gateways along the lane, 975044, the walker stands at the edge of the main Culm outcrop. Behind him, its surface lies at 750-850 feet. Below, the braided course of the river Culm and the meandering Exe flow south across a flat, open landscape of New Red Sandstones (Figure 52). To

FIG 52

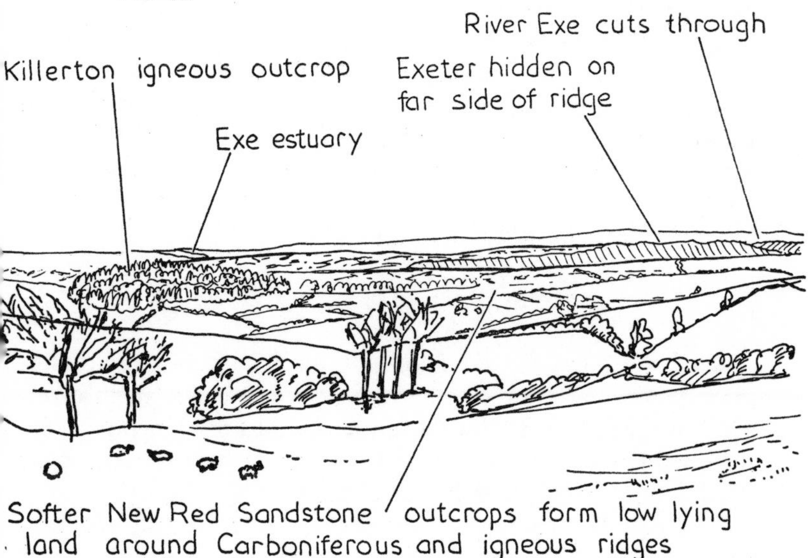

VIEW FROM YARDS DOWNS 975044 S

FIG 53

RELATIONSHIP OF INLIERS & OUTLIERS

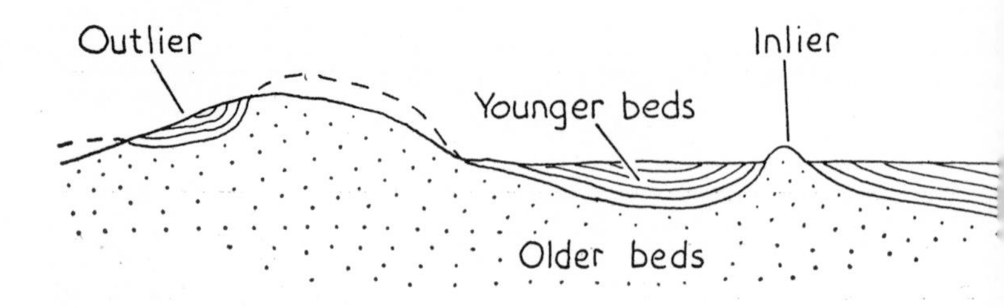

the south-west Culm beds appear again in the Stoke Hill ridge, deflecting the river Exe westwards. It pushes against them in a fine series of river cliffs before eventually finding its way through.

Eastwards the Culm ridge disappears beneath the New Red Sandstones at Poltimore, coming up again as an inlier in Ashclyst Forest to the SSE of our viewpoint. As Figure 53 shows, an inlier is a mass of older rocks reappearing up through a cover of newer materials. Ashclyst Forest and the Stoke Hill ridge, then, are old Carboniferous hills, once completely buried by the New Red Sandstones and now being re-exposed as the weak cover is removed around them.

There is only one other break in the flat land below, the tree-crowned Dolbury Hill with Killerton Gardens. Killerton is one of a series of lava outcrops in the Permian beds around Exeter (see Haldon also, page 159). They have an important history as local building materials.

THE EXETER VOLCANIC OUTCROPS

Bedded among the sandstones, these lavas are remnants of more extensive flows rather than volcanic cones themselves. The only small vents found have been at Sprydoncote in the Ashclyst Forest area, at Crossmead west of the city, and near Webberton Cross below the Haldons.

Dark, reddish coloured lavas are found at Ide and Webberton, and lighter coloured outcrops at Pocombe and Rougemont. Killerton Park provides a third type of light grey colour.

Unfortunately, since the days of the Geological Survey Memoir (1902), many exposures have been covered over. The exposures at School Wood near Webberton Cross are described on page 160 and another reasonable site remains in Crossmead caravan park, 897914, with fragmented basalt sandwiched between coarse breccias and sands. Farther afield

MAP 10

(Generalised from maps of the Institute of Geological Sciences by permission of the Director)

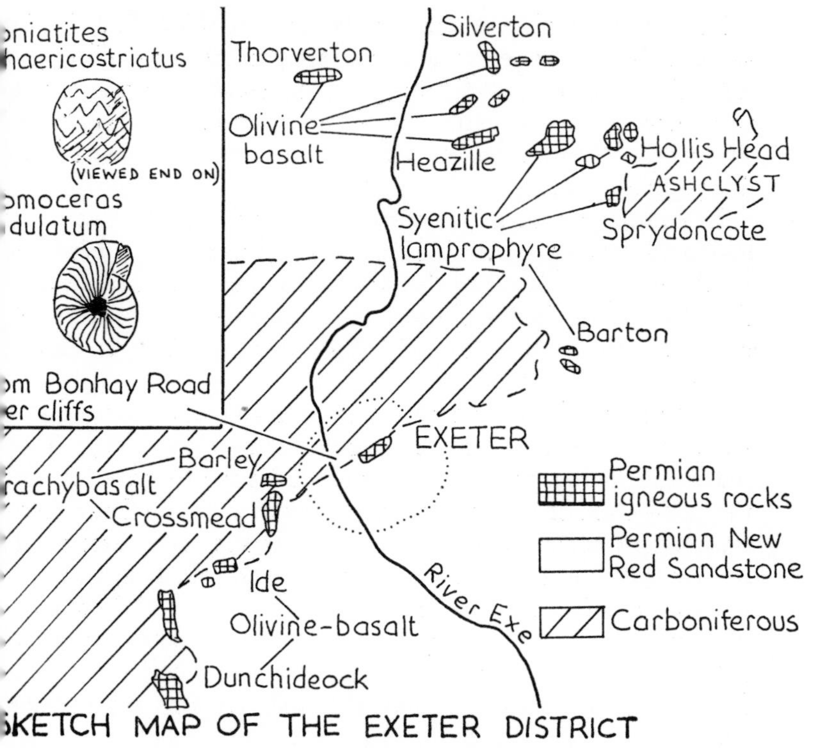

the lavas can be seen at Posbury Clump, south of Crediton, and Hannaborough quarry, near Hatherleigh.

Despite the lack of exposures, the effect of the lavas on the district is obvious in other ways. In some cases they form prominent features—Killerton, the ridge around Old Heazille Farm near Silverton, Rougemont in the heart of the city—but their most notable contribution is as a building stone.

Exeter Castle was built of Rougemont Hill on which it stands. There is the façade of the Royal Albert Museum in

Queen Street where the brownish red stone displays its frequent calcite veins. The stone was also used to face the old city walls—several sections survive around the centre—and in various parish churches of the district, Rewe for example. The highly cleaved Culm shales were obviously no competitor to it.

EXETER'S OTHER BUILDING STONES

Its nearest rivals were the softer textured New Red Sandstone beds and these are also common in Exeter walls, particularly since the old city stood on them and later extensions were eastwards over their ridges. Wonford quarry at Heavitree was worked from 1390, its disused built-over pits now straddling Quarry Lane.

In colour, Exeter is indeed the red heart of red Devon for beside its stones it also has much red brick, many fired in local foundries. Some of the brick pits also worked in New Red Sandstone beds—Clifton Hill athletic track was one, another existed in Polsloe Road—but others are in Culm shale beds.

The principal modern works is in Pinhoe, working these reddened shales from the south side of the Culm ridge. The red colouring stems from the original New Red covering on the hill—from it iron oxide worked down into the Culm in a staining process known as 'raddling'. The shales are excavated from a pit north of the works, 956946, crushed and mixed with water. The mixture is then formed into a stiff column of clay, cut to size and dried for twenty-four hours before it enters a tunnel kiln where the temperature reaches 1010°C. Such works are a far cry from the early days of hand-made bricks.

THE EXE RIVER CLIFFS

Crossing the river on the north side of the city, the Stoke Hill Culm ridge forms an upfolded mass of gritty sands and shale beds. The river Exe has made little progress in widening its valley here despite the added water of its tributaries, the rivers Culm and Creedy. Its achievements so far have been limited to the destruction of any spurs it found and to

FIG 54

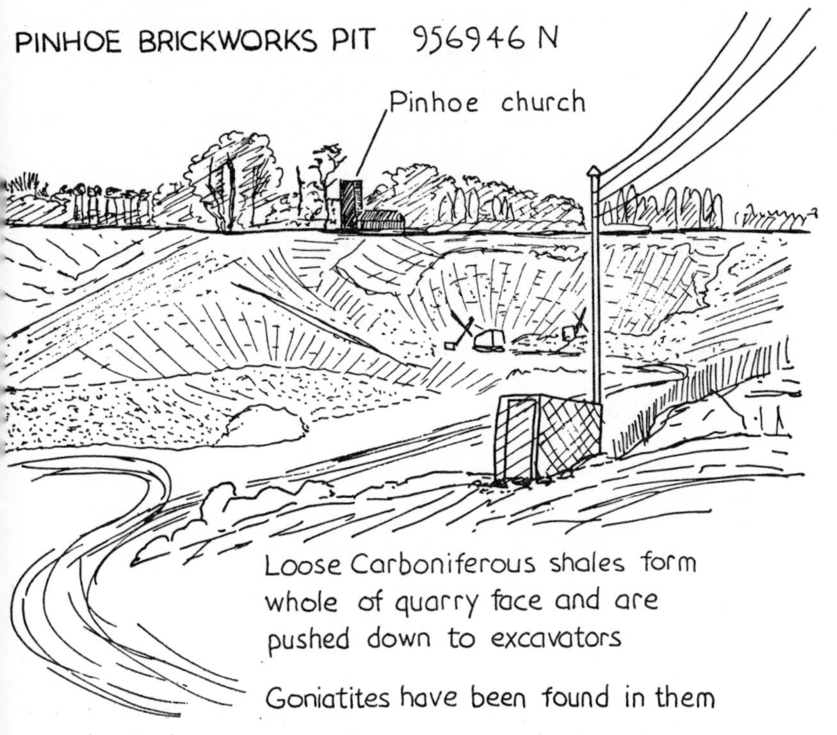

straightening out the sides with a long series of river cliffs.

At times these undermine the main road by Stoke Woods but farther south in Cowley Bridge road and Bonhay road the cliffs stand back from the river now. It is difficult to make much of their geology, rows of houses and walls obstruct the view, but goniatites found at Stoke Woods and near Head Weir prove the beds are Upper Carboniferous (Namurian) in age.

THE SITE OF EXETER

In the sandstone outcrop river cliffs are less noticeable, partly because they would have been less sharp anyway in a softer rock, partly because the bluff which existed has been largely obscured by building. Exe Island at the bottom of the town lay between bluff and river and was a marshy area. Reclaimed at an early date, it became the industrial centre just outside the city wall. From Head Weir the main leat ran around its inner side, following the base of the bluff to provide water power for the industries. The bluff can still be traced around the bottom of High Street and Stepcote Hill.

Two smaller streams break the main valley side at Exeter, the Longbrook and the Shute Brook—they isolate the hill on which the city stands. The Longbrook was particularly valuable, forming a natural ditch outside the northern wall of the city. The steep nature of the valley is amply illustrated by the construction of the North Road Iron Bridge in 1834 and the old steep streets across the brook can still be seen below it. Farther up Longbrook, the walls enclosed the south

FIG 55

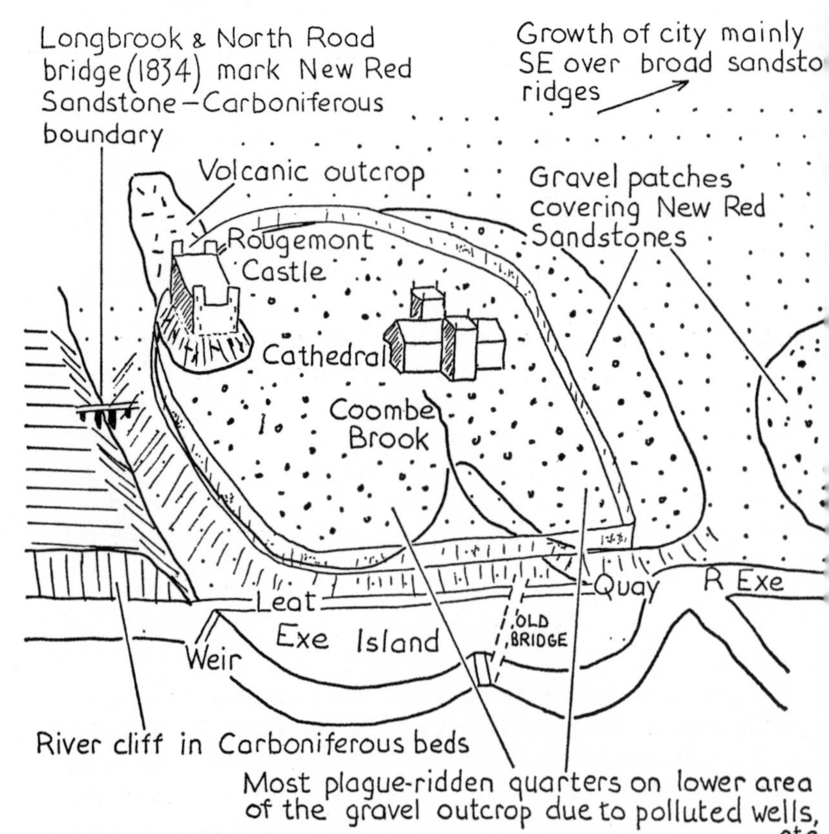

EXETER IN RELATION TO ITS GEOLOGY

end of the Rougemont volcanic outcrop and the castle on its summit.

The southern wall of Exeter makes no use of commanding slopes. It stands on the sandstones ridge, leaving the Shute Brook well outside. The more interesting feature on this side of Exeter is a little stream rising near the cathedral, the Combe Brook. It flows from shallow gravels covering the ridge and its effect on them is easily seen on the Survey's sheet 325, dividing the gravel outcrop into two lobes.

The gravel cover played an important part in the life of old Exeter—it drained quickly after rain, keeping foundations dry. An unhealthy mixture of wells, cesspits and burials could be dug out of it with equal ease! Small wonder that the lower quarters of the town were the most plague-ridden.

THE EXE AND ITS ESTUARY

The river Exe has always been used as a classical boundary by geographers—the Tees-Exe line. This is a convenient line to divide Britain into two physical provinces. To the north-west are the Palaeozoic rocks, Devonian, Carboniferous and others, ancient and resistant outcrops forming high ground. To the south-east the geologist enters the younger Mesozoic and Cainozoic formations, softer and less consolidated beds creating lowland areas. Head Weir at Exeter stands right on the line; the differences between the Culm ridge and the softer sandstones is the reason why it was built there.

Geologically the boundary is fixed rather differently. It used to be looked on as marking the base of the Budleigh Salterton Pebble Beds—leaving all the New Red Sandstones to the west in the Permian (Palaeozoic era) and all those to the east in the Triassic (Mesozoic era). But the recent discovery of Triassic plant spores in beds as far west as Exmouth has destroyed that idea entirely. The boundary must lie there or farther west again. Geologists and geographers are approaching an agreement!

Penetrating eight miles inland, the Exe estuary is enclosed by New Red Sandstone beds. The colourful landscape is rolling and varied but in an open, gentle fashion. There is a direct harmony of settlement and land here, the warm red soil reflected in cob walls and tall sandstone church towers.

This is a countryside to savour, perhaps from the vantage points of Haldon before getting to know it more closely.

A great fault runs beneath the estuary and causes the Langstone Breccias of Dawlish Warren to reappear in Exmouth. The fault probably follows the lower Clyst Valley and then the Exe close to the west bank at Starcross. The river

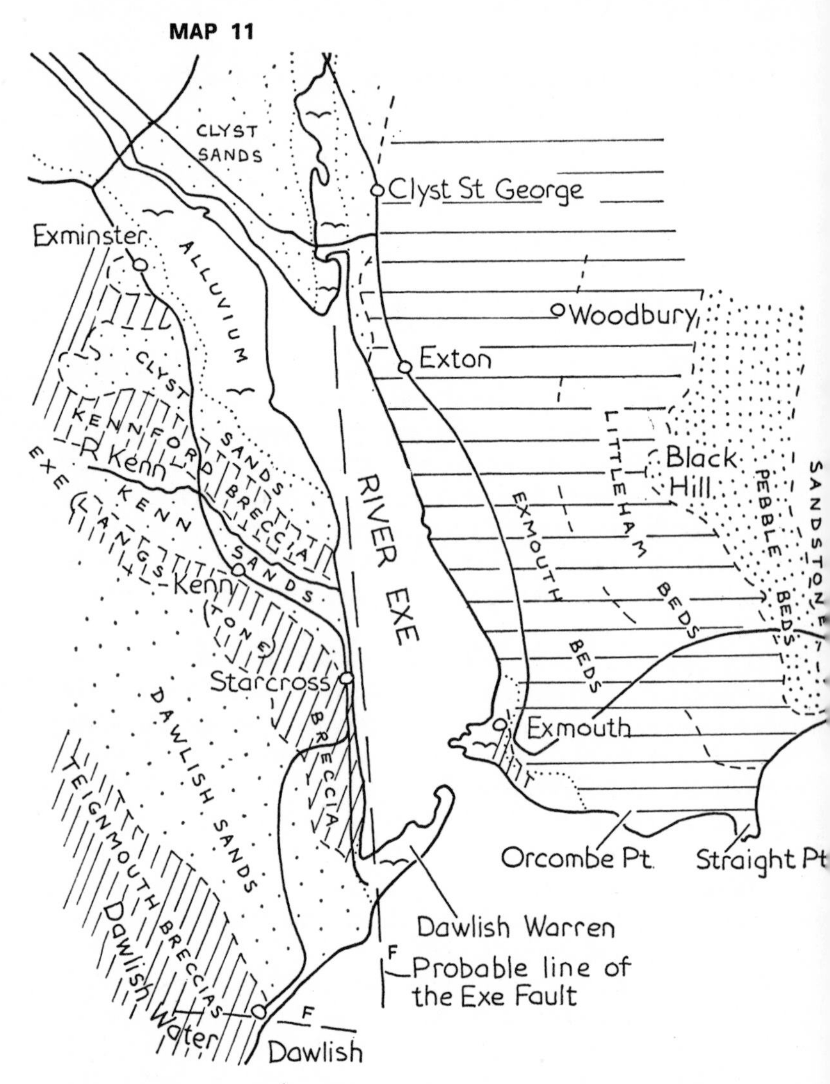

NEW RED OUTCROPS OF THE EXE ESTUARY

obviously found it a useful weakness when creating its estuary.

The fault is a normal one, with a movement estimated between 700 and 1,100 feet. An exact figure would be difficult; the rocks on each side have no useful fossil horizons and the whole feature is well hidden beneath the estuary muds.

Since the estuary's position was well out in the old arid basin, the beds around it are very sandy. Along the west shore the alternation of sands and breccias is clearly reflected in the scenery and can be followed along the Exminster-Dawlish Warren road.

The journey begins in the bright-orange Clyst Sands. On the south side of the Exeter by-pass they provide one of the best sections of wind-blown sands in the district in Bishops Clyst sand quarry. There they are quarried for builders' needs and the bright-coloured heaps can be seen at sites all over the area. At Exminster they appear in the roadside opposite Exe Vale Hospital entrance.

The road rises then onto the Kennford Breccia ridge, crossing its crest at Powderham Arch. Descending the wooded slopes to the south, it returns again to sandstones. There it meets the soft Kenn Sands surrounding Kenton village; the river Kenn has cut its valley along their outcrop. The road also follows them until, near the bank of the Exe, it runs onto the Exe Breccias for the remainder of its journey to Dawlish Warren. The whole route is a very gentle version of a scarp and vale landscape.

One of the best flood-plains in Devon can be studied along the Exe between Exeter and Powderham, and in the lower Clyst Valley. Extensive flats occur, built of brown silt deposits dumped here by the rivers as they lose all gradient on reaching sea level. Along the inner margins of the flood-plain the boundary of silts and New Red Sandstone soils can often be determined within a foot or two when fields have been ploughed. Some allowance must, of course, be made for soil creep; the true junction is buried slightly up-slope.

The Exe still poses flood problems and the protective levels along its banks are seldom sufficient, the flats partly underwater most winters and flood-depth posts necessary where the Topsham-Exmouth road crosses the Clyst Valley.

For features associated with the mouth of the river Exe, see page 115.

CHAPTER 9

East Devon - 1: Exmouth to Sidmouth

Exmouth developed its holiday trade in the early nineteenth century, and prints of the period show the most prominent rows of houses along the front in those days to have been The Beacon and Louisa Terrace. Those houses stood on the top of an old cliff feature just east of the town centre, an outcrop of Langstone Breccias now forming public gardens behind the Imperial Hotel. From the terraces there is an exceptional view of the New Red Sandstone beds, culminating in the old Devonian ridges of Torquay to the south and capped by the flat gravel cover of the Haldon Hills on the opposite side of the Exe estuary. The Exe Breccia cliff is now well inland, its scale more in keeping with the estuary cliffs than the higher ones of the coast.

Small developments of sand dunes along the Maer at Exmouth recall Dawlish Warren opposite. The sand here was blown across the river mouth by westerly winds. Ships leaving the Exe illustrate how the deep-water channel keeps to the east until it is well beyond Exmouth sea front. The Pole Sand on the other side of the channel rests on New Red Sandstone beds where these appear briefly in the Checkstones reef.

The main New Red Sandstone formations in Exmouth are the Exmouth and Littleham Beds. Within the town the Exmouth Beds have been worked for brickmaking. The last active works stands by the main road in Withycombe Raleigh and another existed north of Brixham Bridge. Despite their generally silty character, the soft Exmouth beds create a varied outline along the coast between the promenade and Littleham Cove. The irregularities of the shore are due to thick sandstone beds. More resistant to the waves, they form the headlands of Rodney and Orcombe and the double promontory of Straight Point. At each sandstones either dip down to sea level or have been brought down by N-S faults.

FIG 56

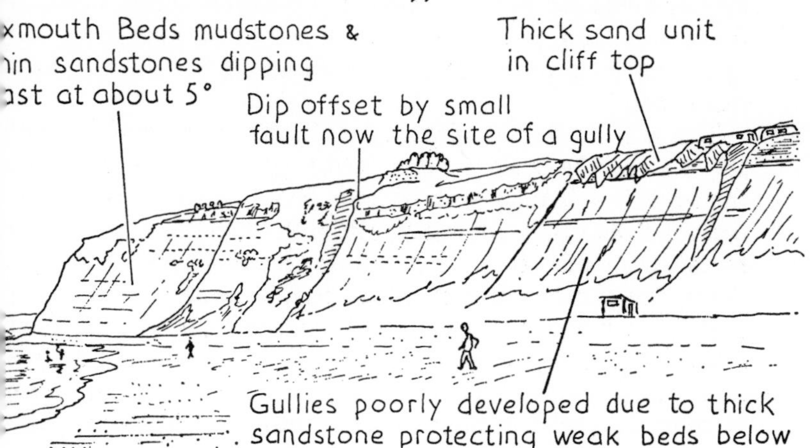

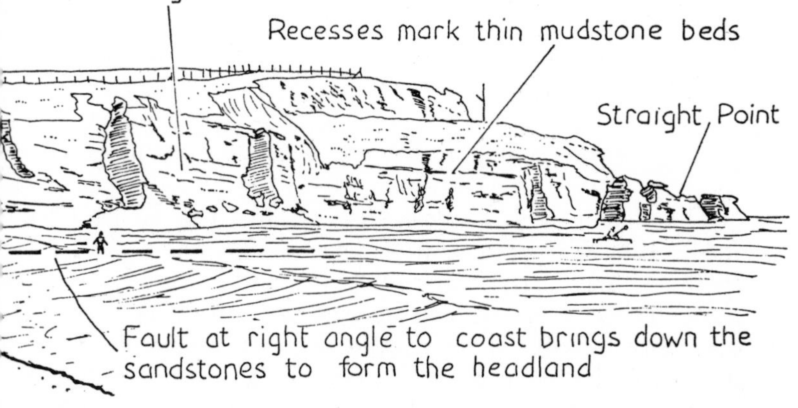

Rodney Point marks the end of the Marine Drive. The small retaining wall behind the road aptly illustrates the weak nature of New Red material—the blocks have been worn back so much that the cement between them now stands out as prominent ribs. At the end of the road a landslip which last moved in 1961 marks the site of faults. Split surfacing on the zig-zag path which climbs this cliff shows that the slope is still unstable.

At low tide the sandstone beds can be traced around Rodney Point to Orcombe Point, east of which they dip below sea level. There are two more faults in the cliffs between the points and near them marly blocks falling from the upper faces have sometimes produced a few fossil plant fragments.

The walker should next make for the coastguard lookout above Orcombe Point. From it the gentle eastward dip of the New Red Sandstone beds, 5-10°, can be traced along the back of Sandy Bay. Several small faults off-setting the dip are revealed by thin sand beds within the mudstones. Straight Point, the eastern limit of Sandy Bay, is another thick sandstone unit brought down by faulting. On a larger scale than Orcombe Point, its sands show cross-bedding and include some thin mudstone bands. These sand beds were the work of river channels running well out into the arid basin while the mudstones may represent brief intervals of lake or backwater conditions.

From Straight Point to Budleigh Salterton there is little change in the nature of the cliffs. The most notable feature is the beginning of the famous Pebble Bed outcrop at West Down dipping gradually down to sea level by the time Budleigh itself is reached.

THE BUDLEIGH SALTERTON PEBBLE BEDS

The Pebble Beds of East Devon reveal just how intimate the relationship between rock type and landscape can be, for there are few parts of the county where the two are in closer harmony. It is always easy to contrast broad divisions—the acid soils of the high granite districts, the often poorly-drained Culm plateaus of north-west Devon, the rich grasslands of marl country. Here geology and land-use are so perfectly linked that observant eyes can place the limits of the Pebble Beds within a few yards.

Similar close relationships occur in the Cretaceous outcrops of East Devon, to be described later, and it is these direct associations which make East Devon so distinctive—coupled, of course, with the very rich fossil record available once the geologist leaves the New Red Sandstone districts. In older Palaeozoic Devon things are more blurred. There is no distinguishing quality in the South Hams landscape to help the observer separate the slates of one part from the volcanics of another. In some ways the geologist may be much happier in East Devon!

Although the Pebble Beds are locally only 80-103 feet thick, their E-W outcrop is often over a mile wide. The reason for this is the very gentle eastward dip, allowing them to remain at the surface long enough to make a distinctive contribution to the district before they pass beneath younger and higher formations to the east.

FIG 57

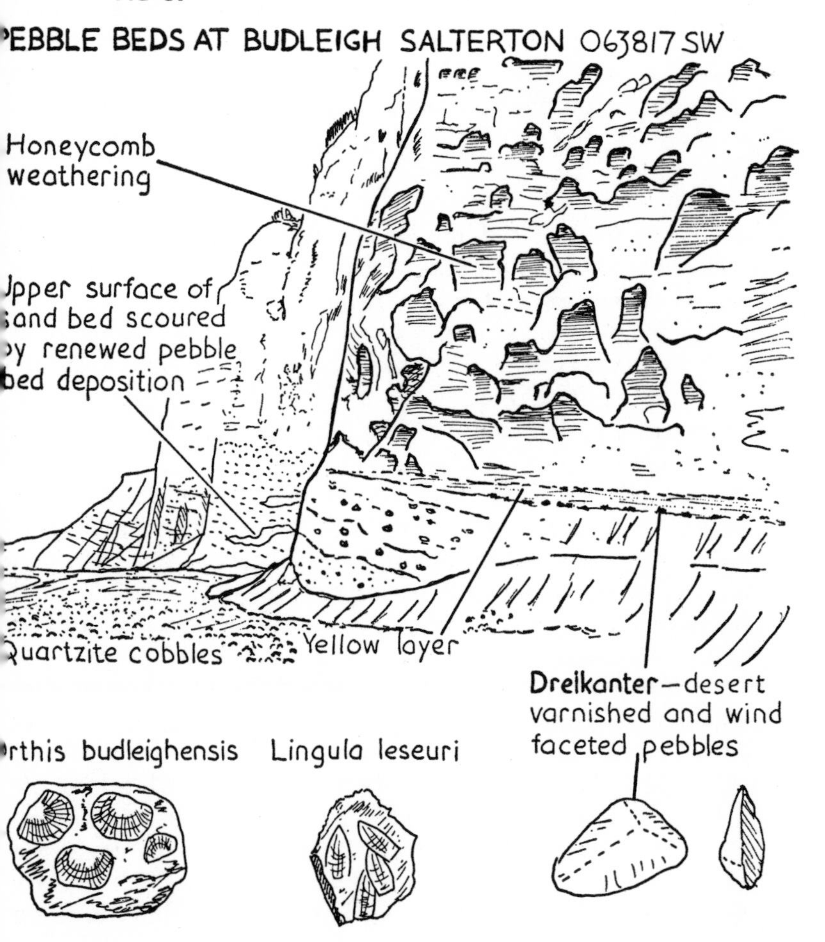

Budleigh Salterton lies in a small valley cut along a fault line marking the eastern limit of the Pebble Bed outcrop. The beds are well-exposed in the cliffs near the promenade (Figure 57). The first exposures reached are marked at the top by a bright yellowish band and a thin layer of blackened pebbles. The yellowish colouring may be due to the water-logged base of a sand dune in the beds which cover them. Alternatively, the normal ferric iron of the red sands could have been changed to ferrous by organic activity. Like the sand-dune origin of the beds above, the blackened pebbles are also evidence that wind action followed the deposition of the Pebble Beds. These black pebbles are dreikanter, wind-faceted and polished stones. Good ones have a triangular appearance with three facets on the top (see inset Figure 57). The black colour is only a desert varnish—split open they are ordinary quartzites like the Pebble Beds below.

There is plenty of evidence of modern wind action in these cliffs. The sand beds have a honeycomb appearance, with hundreds of rounded hollows formed on the faces. Each one is the work of eddying air currents bombarding the cliffs with loose grains and wearing away those parts which are less well-cemented or more easily dried out.

There are also some thick sands within the Pebble Beds and often the tops of those sand beds were scoured away by the river as fresh pebble accumulation began. Pebble-filled channels can be seen which definitely cut into the sand structures.

Although the Pebble Beds are mainly quartzites there are pieces of dark red and grey sandstones. A very small percentage of these sandstones contain fossils. Careful selection of hard pieces with good bedding in them and visible shell fragments is needed but even then the chances of finding *Orthis budleighensis* or *Lingula leseuri* are small (Figure 57). Both fossils are of Ordovician age and the next nearest sites where they can be found now are in Brittany—in the Gres de May and the Gres Armoricain. This seems to prove that the Pebble Beds were deposited by a river flowing into the arid area from the south-west, long before there was an English Channel. A river from Brittany would have needed to be very powerful though and have a good gradient to carry the stones as far as East Devon. But there may have been an alternative source in those days, perhaps a nearer ridge of Ordovician rocks in the Channel area.

A third pebble type commonly found are those bearing tourmaline. A feature of the granite outcrops, they must have been transported eastwards from those areas. The Pebble Bed outcrop continues northwards into Somerset. Away from East Devon its material becomes more local in origin, mainly Devonian and Carboniferous fragments. A few pebbles from these periods can also be found at Budleigh Salterton.

Several small faults in the cliffs are emphasised by jumps in the Pebble Bed horizons and farther west current bedding can be seen in the sandstones below. Occasionally agates can be found, while careful exploration along the water-line can yield semi-precious stones such as carnelian. This is a clear chalcedony, a waxy smooth form of quartz; it is generally red or reddish brown in colour.

East of Budleigh Salterton the cliffs are formed in Otter Sandstones. The Otter Valley itself is choked with alluvium and all traces of its estuary have disappeared. The river is hardly able to maintain its mouth and the small outlet has been pushed over to the eastern side of the valley by the dominant eastward drift of material along the coast. Otterton Point helps hinder the clearance of the river mouth.

The Otter Sandstones are exposed as far as Sidmouth and cliff erosion in them is much slower than in the marly beds near Exmouth. In places, isolated pinnacles have survived for the time as sea-stacks—there are some to be seen in Ladram Bay.

THE EAST DEVON COMMONS

West Down, East Budleigh, Bicton, Woodbury, Colaton Raleigh, Aylesbeare—the roll-call of the villages below the Pebble Beds marks the sinuous outcrop of commons running north from the coast. The rich marl pastures end abruptly where the Pebble Beds appear and a land of pine woods, heather, bramble and rough grass begins. The mixture of colours, greens, purples and withered yellow is delightful.

Only an occasional stream is seen, surviving because it has cut down to the marls below—the one flowing down to Budleigh Salterton, for example. With its upper areas now in marls it has proved useful for two reservoir sites and drainage from the surrounding gravels is collected here for the Exmouth supply. Water can usually be seen trickling across the road between the two reservoirs, near Four Beeches. In other marl-floored valleys on the eastward dip-slope, conditions are damp enough to support small valley bogs with carpets of sphagnum.

There is a similar contrast in the woodland. The pines on

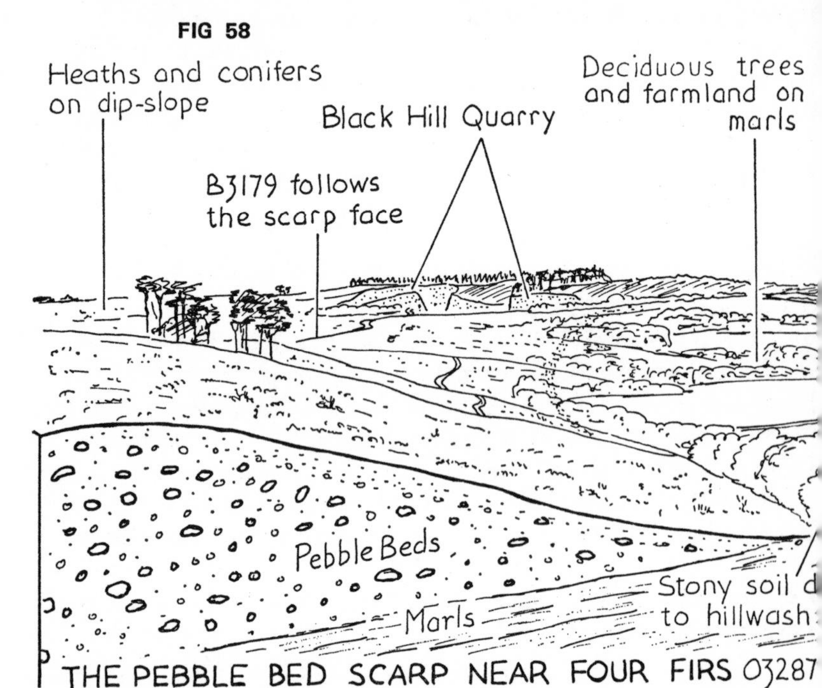

THE PEBBLE BED SCARP NEAR FOUR FIRS 03287

the heaths, introduced into the area in the late eighteenth century, look thoroughly natural features now. As soon as marl ground is reached to east or west the woodland becomes more varied, a mixture of deciduous and coniferous. There are many beech trees in these lower woods.

The scarp form of the Pebble Beds is best seen on the B3179 from Knowle to Four Firs and then along the B3180. At times the road runs at the base of the scarp, at others along its crest. The many small roadside gravel pits are a feature of all the commons and, like the large working quarry at Black Hill, provided road aggregates.

Black Hill quarry on the north side of Lympstone Common is the biggest inland exposure of Pebble Beds. The section reaches 103 feet, a thick mass of cobbles buried in bright-red sands. The cobbles are crushed and screened to a variety of sizes. Development of the quarry is a problem, standing as it does in such a beautiful area. Preservation of pine trees explains the untouched ridge projecting from its face.

There is a car park on the crest of the scarp just north of the Four Firs junction. Figure 58 shows the view back to the quarry. Pine woods rise above it, while the scarp face below is clothed with heathland plants. Abrupt changes occur at the base where the marls are revealed by the pasture lands and orderly hedgerows. In places, a good deal of gravel has been washed down the scarp-face making the marls rather stony. The farmlands to the east of Woodbury and Aylesbeare are affected by this.

At the old hill fort of Woodbury Castle the scarp reaches 560 feet in height. From the castle there is a breath-taking panorama embracing the whole New Red Sandstone landscape around the Exe estuary. To the west, the Haldons dominate the skyline and glimpses of Dartmoor hills beyond reveal the 'old lands'. Northwards, Exeter is backed by the Culm hills, rising from beneath the New Red beds—the University estate can be picked out on their slopes.

Leaving Woodbury Castle, about one mile farther north in the car park opposite the Warren there is another fine view —this time of the high ridge line overlooking the Otter Valley. The ridge follows the river from Bowd to Gittisham. At the distant end, Honiton can be sighted with the distinctive tree clump on Dumpdon Hill beyond. This ridge is capped by the great unconformity of East Devon and study of it is best commenced by returning to the coast at Sidmouth. This is the second major unconformity in Devon and Figure 59 shows the effects this one and the one at the base of the New Red Sandstones have on the strike of the South Devon rocks. In the Devonian and Carboniferous areas the rocks strike E-W. In the New Red there is a change to N-S, while in the Cretaceous the beds have a flatter attitude, forming the great structural plateaus of East Devon.

FIG 59

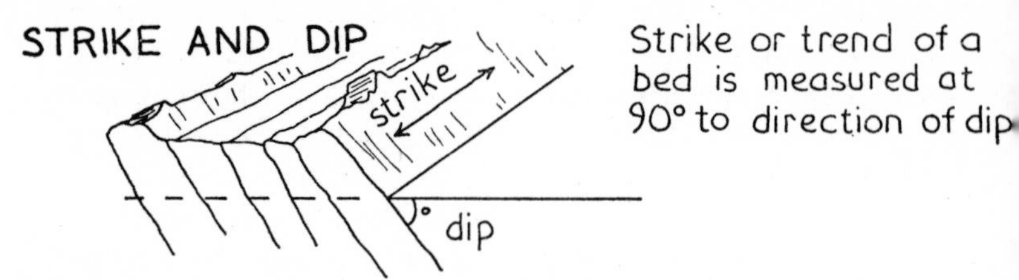

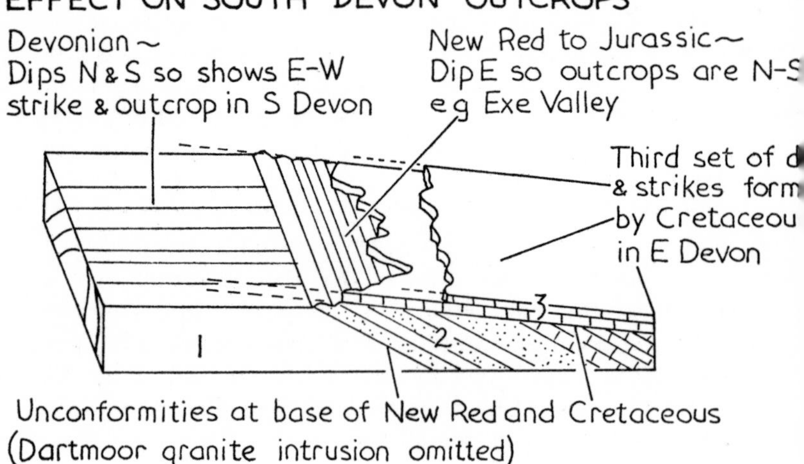

THE GREAT CRETACEOUS UNCONFORMITY

From the clifftop above Connaught Gardens, Sidmouth, the views eastward to Beer Head and inland along the Sid Valley reveal the effect of the unconformity on the East Devon landscape. Figure 60 shows stages in the formation of the East Devon rocks.

In terms of time, Sidmouth lies midway between the Pebble Beds of the early Triassic (Bunter) and the marls which were characterisitc of the later Triassic (Keuper) times. The names Bunter and Keuper are taken from Triassic beds in Germany.

Figure 60 shows how the arid basin continued to fill up during Triassic times, the material becoming finer and more dusty, with occasional temporary lakes and salt pans. These

FIG 60

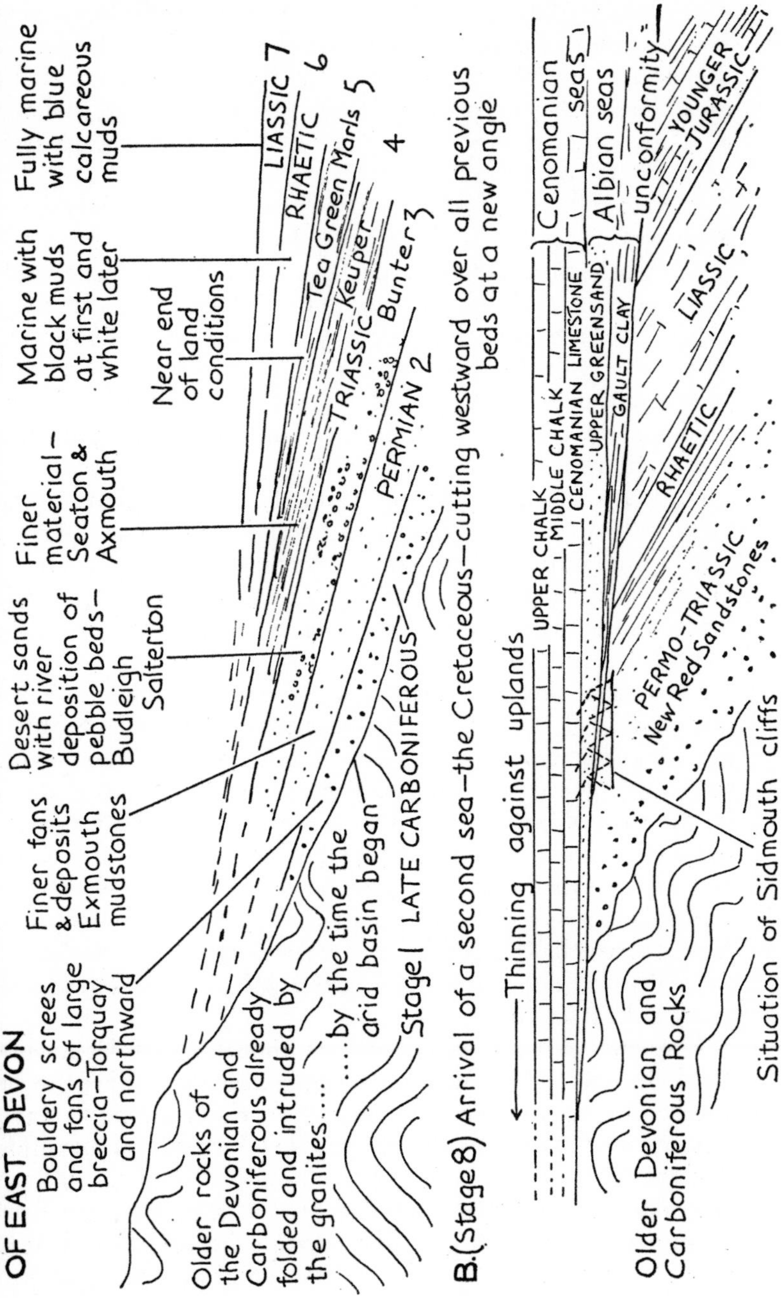

OF EAST DEVON

Bouldery screes and fans of large breccia–Torquay and northward

Finer fans & deposits Exmouth mudstones

Desert sands with river deposition of pebble beds–Budleigh Salterton

Finer material–Seaton & Axmouth

Near end of land conditions

Marine with black muds at first and white later

Fully marine with blue calcareous muds

Older rocks of the Devonian and Carboniferous already folded and intruded by the granites.... ....by the time the arid basin began

Stage 1 LATE CARBONIFEROUS

PERMIAN 2

Bunter 3

TRIASSIC

Keuper

Tea Green Marls 5

4

RHAETIC 6

LIASSIC 7

B. (Stage 8) Arrival of a second sea–the Cretaceous–cutting westward over all previous beds at a new angle

←Thinning against uplands

UPPER CHALK

MIDDLE CHALK

CENOMANIAN LIMESTONE

Cenomanian seas

UPPER GREENSAND

GAULT CLAY

Albian seas

unconformity

Older Devonian and Carboniferous Rocks

PERMO-TRIASSIC New Red Sandstones

RHAETIC

LIASSIC

YOUNGER JURASSIC

Situation of Sidmouth cliffs

beds form the bulk of the cliff-faces east of Sidmouth. Dipping gently they eventually pass below shore level near Branscombe. The rest of the sequence in Figure 60 shows that at the end of Triassic times the Rhaetic sea invaded the area. Gradually deepening, it formed the dark grey shales and limestones seen at Lyme Regis and Portland, the Jurassic beds, 135 to 180 million years old.

In Figure 60, Stage 8, we can see how all this old landscape of Jurassic, Triassic and Permian beds was tilted and submerged beneath the Cretaceous sea. This sea ran westwards over the top of them, creating a new series of nearly horizontal beds and even thinly covering the surface of Dartmoor. These are the beds which make the distinctive greenish and yellowish upper cliffs around Sidmouth—the Upper Greensand and the Clay with Flints. Their base is the great unconformity.

Standing on the road to Peak Hill, the explorer must visualise the coast as a continuous rock face without the river Sid and the other valleys which interrupt it today. The Upper Greensand and the Clay with Flints, ignoring the dip of the Triassic beds below, would pass across from ridge to ridge as a continuous sheet.

Looking at the one-inch maps 326 and 327, there are two outstanding features which emphasise the unconformity. Noticing the great spread of Cretaceous (green colours) across the map, observe the changing beds revealed beneath them in the valleys. From east to west they pass successively over older and older rocks. Clearly the base of the Cretaceous beds is a time-break, an interval of erosion. The second feature emphasised by the map is the almost plateau-like level of these beds. Because of it, their outcrops closely follow the contours and the various colours of the map emphasise every individual river basin.

The level outcrop is not perfect though—there is a very gentle dip both eastwards and towards the coast. This can be checked by comparing the height of the capping on the western and eastern sides of any local ridge, for example Buckton Hill east of Sidbury (Figure 63). Above the village the base of the Greensand lies at 450 feet, but just over a quarter of a mile east, on the opposite side of the hill, it is about 20 feet lower. On the coast it descends from 500 feet near Mutters Moor to sea level at Branscombe Mouth, seven miles to the east.

Because of the slight dip, underground drainage in the ridges emerges as springs on the east slopes. These slopes have looping outlines of Cretaceous beds, curving round the many small valleys. Only the valley heads above the springs remain within the Greensand outcrop and these steep-sided, narrow combes are known locally as 'goyles'. In contrast, the Cretaceous outcrops on the western slopes are straight—East Hill overlooking the Otter Valley is the best example.

Exploring inland from Sidmouth, the walker can make his geological map as he goes along. Heavy farming soils on the Keuper marls extend up the valley sides. Then, in the final climb to the ridge summits, the gradient steepens from the 1 in 10 of the marls to around 1 in 3 on the Greensand and Clay with Flints. Since these rougher hill tops are often planted with conifers, the unconformable beds can also be identified by their woodlands.

PEAK HILL, SIDMOUTH

Connaught Gardens and Chit Rocks are formed of sandstones from the base of the Keuper. They have been brought up along a fault which reaches the coast just west of the gardens and form the only interruption along the shore between High Peak and Branscombe. The geologist should study the smooth outline of this coast before making for the summit of Peak Hill. Larger valleys like the river Sid's reach sea level but smaller ones have been left higher up in the cliffs by the speedy erosion of the soft marls. The end of each ridge has been cut straight through, leaving broad triangular-shaped cliffs.

Where the road bends inland above Peak House a cliff path can be followed. It provides views of the rubbly Upper Greensand and Clay with Flints beds both in Peak Hill and in the detached patch capping High Peak farther west. The Greensand on Peak Hill is about 60 feet thick and, paradoxically, grey in colour, like the Greensand of the East Hill ridge and of the Blackdown Hills to the north of Honiton. Farther east the Greensand is more calcareous (Page 153) and divides into the Chert Beds and the Foxmould.

FIG 61

THE COAST EAST OF SIDMOUTH 117870 E

Farmland on New Red Sandstones
Woods, heaths and steeper slopes on nearly level Greensand capping
Beer Head and Hooken landsli
Sidmouth
Eastward dip of New Red beds seen in triangular cliffs at ends of ridges

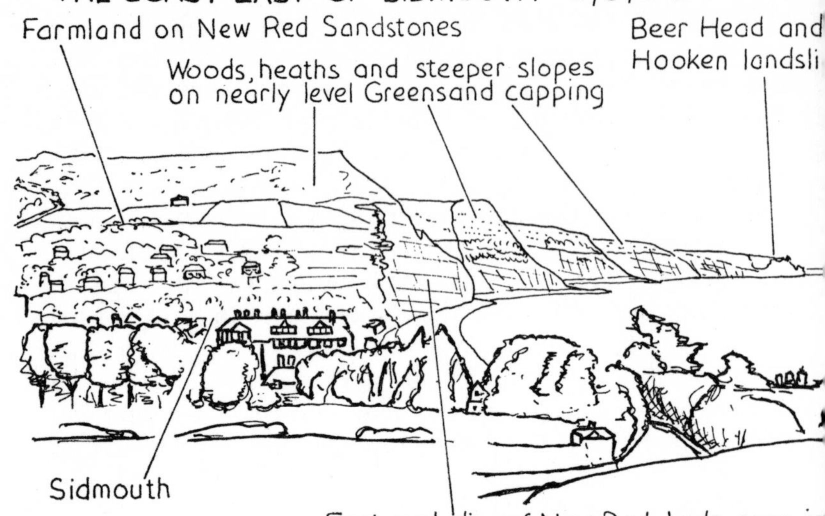

FIG 62

CRETACEOUS COVER FROM PEAK HILL 111868 SW

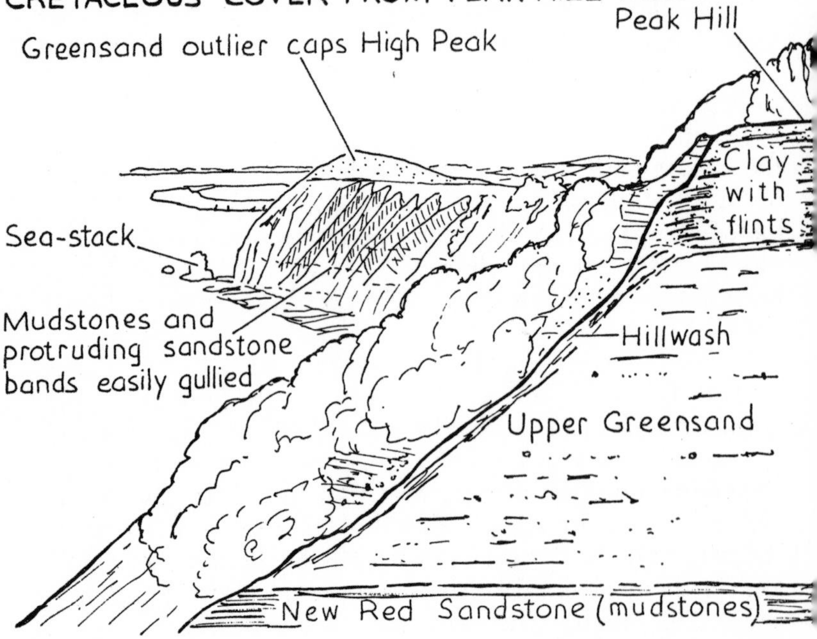
Peak Hill
Greensand outlier caps High Peak
Clay with flints
Sea-stack
Mudstones and protruding sandstone bands easily gullied
Hillwash
Upper Greensand
New Red Sandstone (mudstones)

If, however, the explorer keeps to the road above Peak House he will see how, in common with so many others in the district, its change of direction is due to the Greensand. The road runs diagonally across it in an effort to overcome the steeper gradient. The roads up Salcombe Hill east of Sidmouth, Trow Hill east of Sidford on the A35 or any lanes up the west side of East Hill, provide further examples. Each manages the marl slopes at right angles to the contours but turns diagonally across the Greensand.

The top of Peak Hill provides one of the best distant views of the Pebble Bed outcrop. A dull blue-green line of pine commons to the west interrupts the fertile marl and sandstone country. Near at hand is the Greensand capping covering High Peak and Mutters Moor. It appears again in the far distance at Haldon Hills, but there it is itself covered by younger Eocene gravels instead of the Clay with Flints (see Chapter 11).

FIG 63

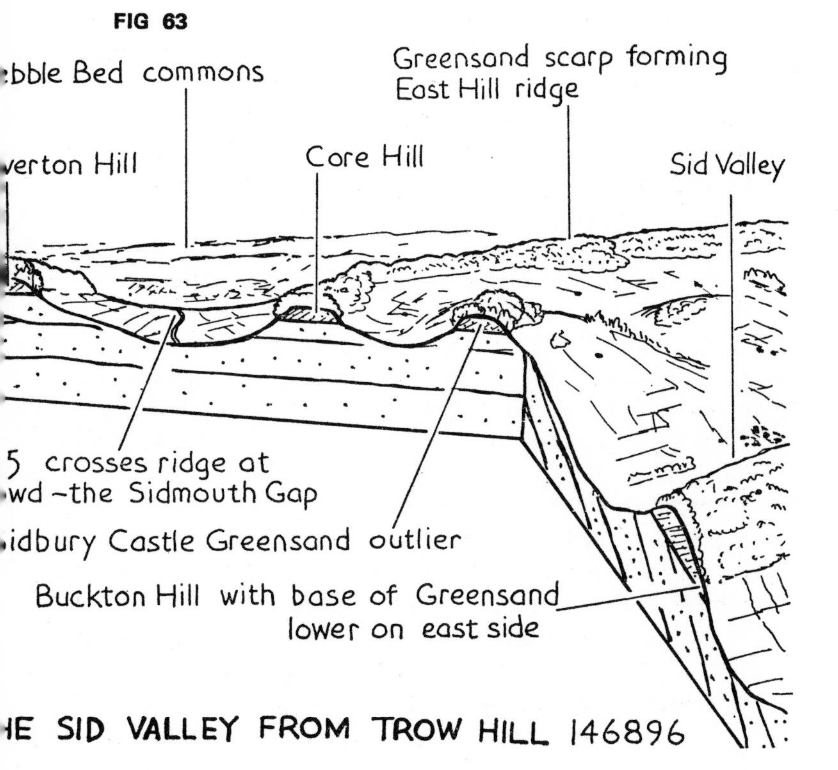

Leaving the Sidmouth district, there is a worthwhile stop at the top of Trow Hill above Sidford. The whole excavation of the Sid Valley down through to the marls can be appreciated here. Cap rocks are obvious on every hill-top with Buckton Hill the most graphic example. In places its higher slopes were too steep to remain stable and landslips of gravels have occurred, spreading down onto the marls. Across the river, part of the capping on the dip-slope of East Hill became isolated and made a convenient defensive site for the Iron Age fort of Sidbury Castle. Westwards, the A35 makes use of the only significant break in East Hill ridge, the Sidmouth Gap at Bowd.

SETTLEMENT IN EAST DEVON

East Devon is an area of contrasts. There are the dry, hungry soils of the Pebble Beds and gravelly Greensands on the one hand, and the rich marls on the other. Among their more obvious effects are the differences in land-use, the variations in the plant life of valleys, heaths or hedgerows already mentioned. There are more subtle influences than these however and the OS one-inch map reveals, for example, a close bond between geology and man's social organisation. Often the very spot on which a house was built was determined by geology.

The paramount need for water led to the siting of farms and villages below points where streams emerged beneath the gravels. While the East Devon settlements are not quite such classical examples of this scarp, spring-line and settlement relationship as parts of south-east England, they do come a close second. Figure 64 represents the usual relationship. Examples are found all over East Devon and this might easily be a diagram of East Hill, with its string of farms below the western face, each on its own stream.

Examples are not limited to areas with straight scarp edges however. Erosion of the cap rocks in East Devon has created almost enclosed basins in some areas—eg, Farway, south of Honiton. The church and its cluster of houses is central to almost every farm in the parish. Each in its own combe, the farms are all within half a mile of the village.

On the south-west margin of the Blackdown Hills over-

FIG 64

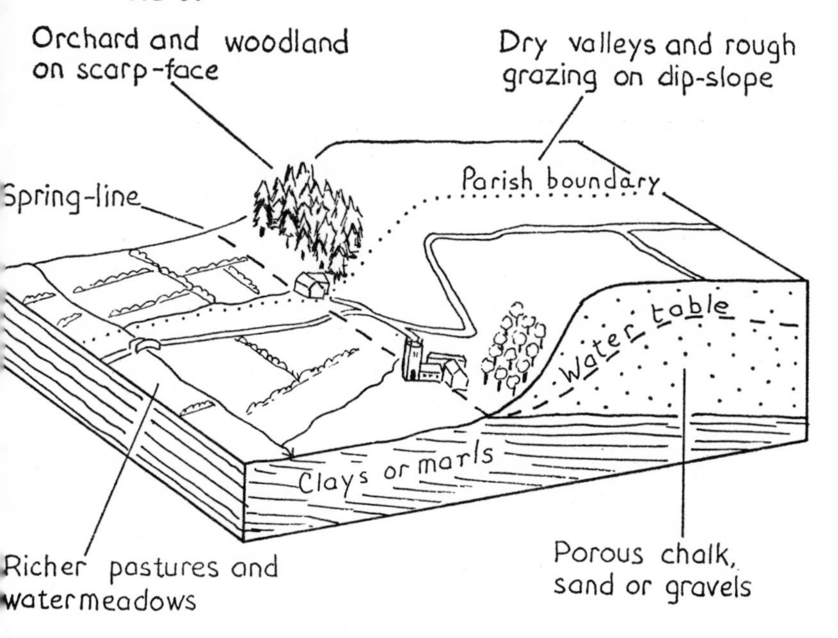

SCARP & SETTLEMENT

looking the Honiton-Cullompton road the edge of the high ground is very irregular but the same close relationship exists. Combe Raleigh, Awliscombe, Broadhembury and Kentisbeare each reveal the wisdom of the old parish boundaries. They were drawn to ensure that every village had its share of the good and bad land. Each has a common up on the cap rocks, some steep intermediate slopes so useful for orchards, and richer meadows on the marls below. Similar boundary patterns can be traced around the Pebble Bed outcrop.

Chapter 10

East Devon - 2: Beer to Lyme Regis

Beer is an attractive village enclosed in a short steep-sided valley, with the stream still flowing down the main street. The beach is mainly flint and quartzite, like most East Devon beaches, but the cliffs are creamy white, rising through gently folded beds to pinnacle forms at the top. These are the most westerly chalk cliffs in England.

There is a good deal of flint cobble in the buildings and much of the beautiful golden-coloured Beer Stone. The public shelter on the east cliff path is built of this stone which comes from a particular horizon near the base of the Middle Chalk beds.

THE BEER STONE QUARRIES

Beer Stone is a gritty, calcareous freestone. It is soft and easily cut but soon hardens when exposed to the air, qualities which have been known since Roman times at least. The exciting thing about the Beer quarries is that they were worked by underground tunnels which run north and south of the road, about three-quarters of a mile to the west of the village.

The stone was used in Roman villas and public buildings, and because of the quarries' convenient position on the coast, it was sent by ship to many cathedrals and parish churches, to St Stephens, Westminster in 1362 and to Rochester in 1367. In Devon, it was used in Exeter's cathedral and guildhall and in the white colonnades and screens of many parish churches. In Beer parish, the Tudor-aged Bovey House and many cottages are built of it.

The dark tunnels which produced the lovely golden stone are historic records themselves. The main workings lie south of the road beginning with Roman diggings where tool marks can still be seen on the walls. The excavations continue into the area worked by the Normans—almost creating an underground cathedral—and beyond are chambers worked at later periods. One tunnel eventually emerges in the cliffs at Hooken Landslip (see below).

FIG 65

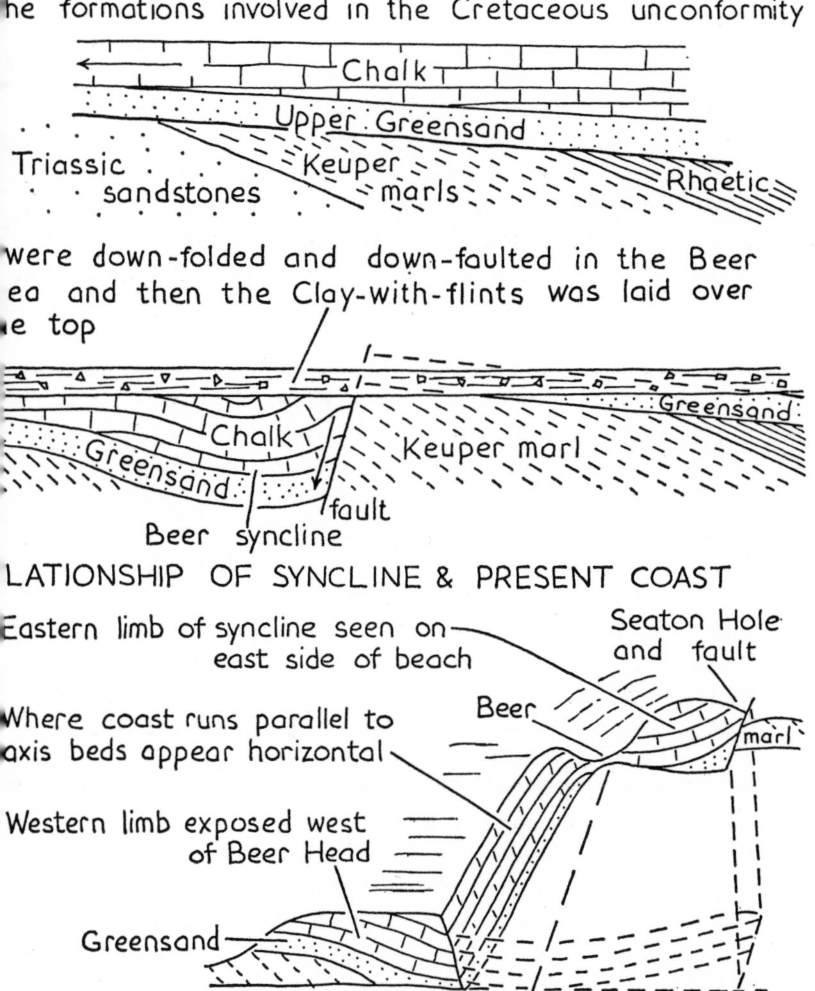

North of the road the quarries are still active, working the chalk above the Beer Stone for burning as agricultural lime. There are some shorter tunnels here but they are not used since the small amount of Beer Stone needed today can be obtained from the quarry floor after the chalk has been stripped off. In the western face the higher chalk is riddled with deep solution pipes. The work of descending streams, the pipes became filled with sand and gravel, giving the whole face there a more dirty appearance than usual.

THE MOST WESTERLY CHALK CLIFFS IN ENGLAND

Walking eastward from Sidmouth, the chalk beds appear first at Maynards Cliff, lying between the Greensand and the Clay with Flints. Their thickness remains fairly constant along the gentle eastward dip as far as Branscombe Mouth, but beyond that point they increase rapidly to form the entire cliff face around Beer and as far east as Seaton beach.

The Greensand beneath almost disappears, running at sea level near Beer beach, and the Keuper Marls pass below sea level and are not seen again until they outcrop on the far side of the fault at Seaton Hole (Figure 65).

Downward movement on the west side of this fault is one reason for the thick chalk development at Beer, but the district is obviously down-folded as well (Figure 65 a and b). From the centre of Beer beach the saucer-like effect is easily seen, beds in the cliffs rising gently eastwards and westward. The rise, in fact, seems steeper eastwards where the explorer is looking more directly up the side of the fold than to the west where, because of the direction of the coast, he looks along the fold axis. The real gradient on the west side can be found beyond Beer Head (Figure 65c).

The walker should have a firm idea of this history and of the sequence of beds involved before he sets off to study the cliff sections, if he is to get full enjoyment from them. The sequence of beds represented between Branscombe and Seaton Hole is given below, and their zone fossils are illustrated in Figure 66.

Once the geologist leaves the New Red Sandstones and begins to explore these Cretaceous beds he moves, from a fossil

		Thickness
UPPER CHALK (SENONIAN)	Only seen at Annis Knob E of Beer beach and in cliff top exposures on Beer Head *Micraster cortestudinarium and Holaster planus*	90ft
MIDDLE CHALK (TURONIAN)	Upper part: *Terebratulina lata* with many flints and some flint-free bands Lower part: *Inoceramus labiatus,* flints rare, nodules of chalk pebbles often present; Beer Stone at the base in places Division C: only found occasionally Soft chalky beds with orbyrhinchids	110-235ft
	INTERVAL OF EROSION	
CENOMANIAN LIMESTONE	The local equivalent to the Lower Chalk of SE England Division B: hard, white limestone with quartz grains and the green mineral glauconite	0-5ft
	INTERVAL OF EROSION	
	Division A: hard, white shelly limestone. Common fossils *Mantelliceras, Hyphoplites*	0-18ft
	INTERVAL OF EROSION	
UPPER GREENSAND AND GAULT	Top Sandstones: fossils rare here	8ft
	Chert Beds: grey or yellowish sands with brown and black cherts and broken oyster shells	70ft
	The Foxmould: green or yellow greensands with glauconite. Oysters, *Exogyra conica,* and scallops, *Serpula concava,* are common Cowstones: elliptical shape fossiliferous lumps	Up to 150ft
	Gault: a few feet of dark loam, only seen locally at Culverhole Point E of Seaton	

viewpoint, from rags to riches. From almost complete lack there is now embarrassing wealth. Terms like 'assemblage' and 'zone fossil' become very important.

The assemblage of fossils is the total population in a particular bed—the frequency of one species compared to

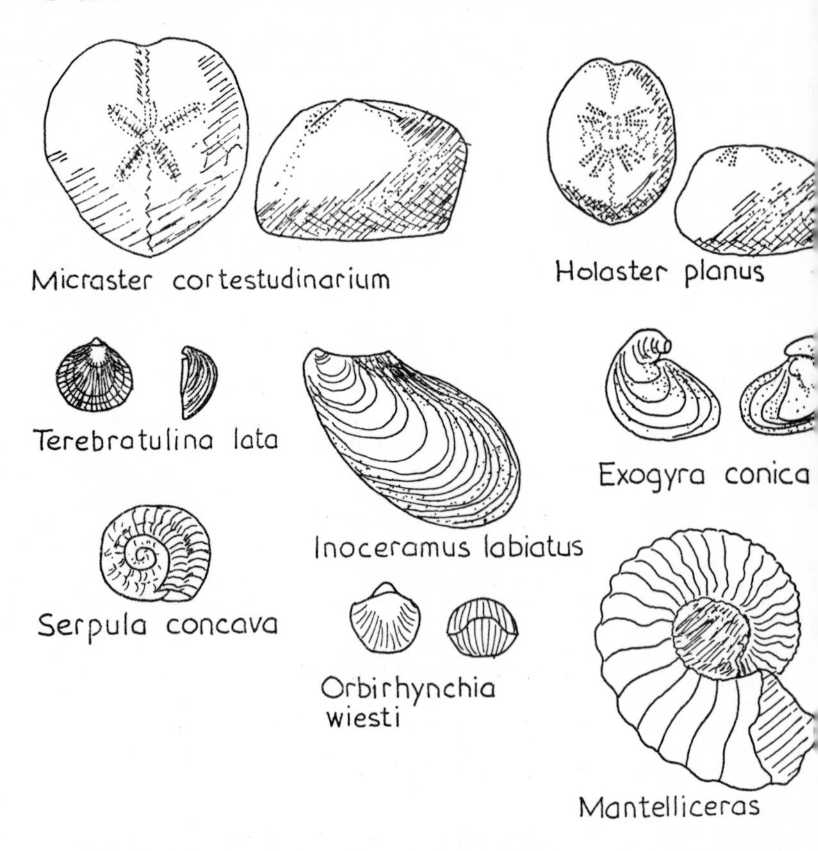

another helps the geologist to place the beds in their correct order of formation. The zone fossil is, in a way, the head of the family, the most important one in the group since it reached its peak development at that time.

HOOKEN LANDSLIP AND BEER HEAD

Parking at the top of the steep hill west from Beer, make for the coastguard lookout on Beer Head. A superb view unfolds below the walker as he reaches the clifftop and the start of the path down to the landslip. Inland the sheer white cliff is broken by the dark mouth of the Beer Stone adit, while in

FIG 67

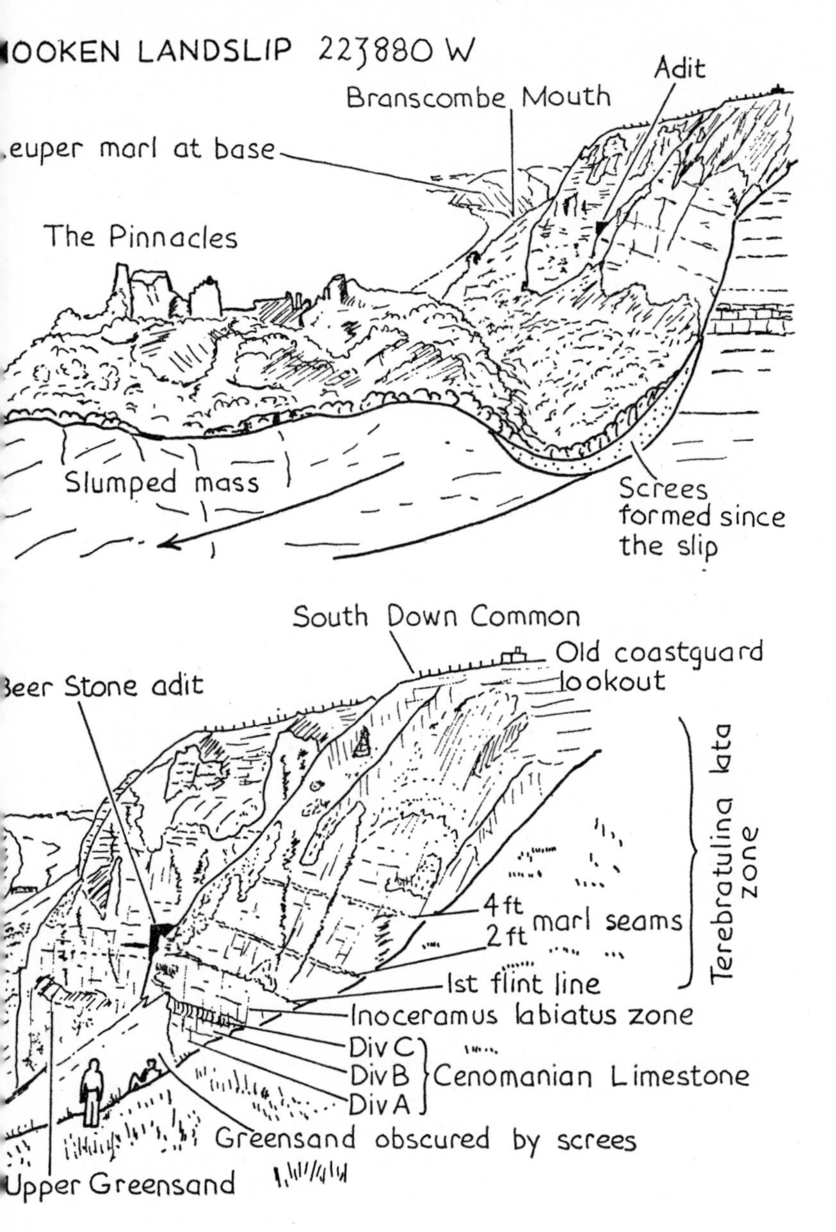

OOKEN LANDSLIP 223880 W
Adit
Branscombe Mouth
euper marl at base
The Pinnacles
Slumped mass
Screes formed since the slip
South Down Common
Old coastguard lookout
Beer Stone adit
Terebratulina lata zone
4 ft
2 ft
marl seams
1st flint line
Inoceramus labiatus zone
Div C
Div B
Div A
Cenomanian Limestone
Greensand obscured by screes
Upper Greensand

the green trough below the path wends its way to Branscombe, the fallen blocks rising between it and the shore as a series of teeth-like pinnacles (Figure 67).

The Hooken Landslip is one of several in East Devon but it is not the largest. Dowlands and Bindon lie farther east and Lyme Regis knows the problem of these great earth-slides well. The movements have usually been quite sudden, their cause the difference between porous chalk and Greensand beds in the clifftops and the more clay-like Keuper marls, Rhaetic and Blue Lias beds below. Underground drainage descends easily until it reaches the top of the marls. There it is forced to spread sideways, making the marl surface into a gigantic slide. Tilted slightly seawards, the surface allows masses of chalk to slip off the cliffs. The cause is simple but the view of Hooken confirms the results as spectacular!

Before the slip occurred, South Down Common ran level to the edge of a sheer cliff. Halfway down its face a stream emerged. About 1788 the stream somehow became blocked underground and must have spread out over the marls within the hill, lubricating their surface. A year later a great fissure appeared in the clifftop cutting off about ten acres of land. Suddenly, one night in March 1790, the isolated land slipped down 250 feet, pushing up a great ridge in the seabed beyond and extending the shore outwards nearly 200 yards. Crab-pots laid 8 to 10 feet below surface the evening before were found 15 feet above water!

The path down into Hooken is open and steep at first but soon begins to weave through thick bushes. Frequent stops are needed to get the best views of the cliff exposures and to avoid missing side turnings to the Pinnacles or the Beer Stone adit.

The Beer Stone adit is a convenient marker since it lies near the bottom of the *Inoceramus labiatus* chalk zone (Figure 67). Immediately below it a niche in the cliff, largely hidden by hill wash, represents the Division C bed. The other divisions of the Cenomanian Limestone follow and the base of the cliff is in Upper Greensand, sometimes hidden behind mounds of fallen waste. The most notable feature here is the total thickness of the Cenomanian Limestone, nearly 24 feet. At Beer beach, described below, it is as little as 3 feet.

Above the adit there is a prominent line of flints—the

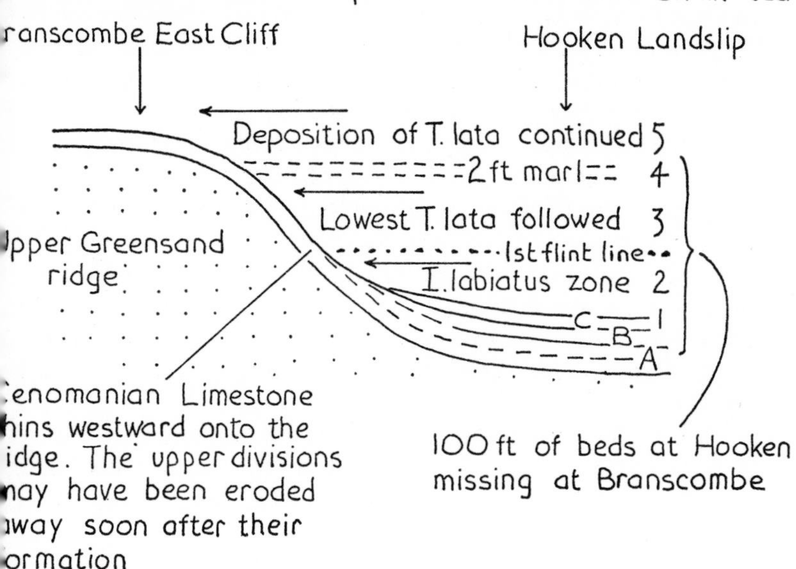

boundary of the Inoceramus zone with the rest of the Middle Chalk, the *Terebratulina lata* zone. Forming the whole upper face these beds are only broken by two thin marly seams.

At the bottom of the slip the path divides before reaching the first of two prominent fallen blocks. Known as Martin's Rock, the block is formed of Greensand beds. This is the point at which to get down onto the beach and look at the western cliffs of Hooken, or walk back to the bottom of Beer Head. If the turning is missed there is no option but to go on past Mitchell's Rock and into Branscombe.

The western cliff at Hooken is inaccessible, so binoculars are useful. Figure 68 illustrates the differences between it and the eastern cliffs in the landslips. The Greensand rises westwards in a bank known as the Branscombe Mouth ridge. The ridge existed in chalk times, for the chalk beds resting against it overlap each other. As they formed, each one extended farther west on to the ridge than its neighbour below. The result is that in Branscombe East Cliff the Terebratulina zone lies directly on top of the Cenomanian Limestone and about 100 feet of chalk strata, present in the landslip, have been eliminated.

Some beds reappear on the far side of the Greensand ridge in Branscombe West Cliff but oddly enough the Inoceramus chalk zone there lacks its own zone fossil—it did not migrate around the ridge into that part of the chalk seas.

The beach below Hooken landslip is part of the unbroken bank of flints and quartzite cobbles running east from Sidmouth. It is usually heavily stepped by storm waves and progress to the base of the Pinnacles is best made if you keep to one of these shelves. Once undermined, the Pinnacles soon fall onto the beach and are broken up, but meanwhile they provide a chance to see some of the beds more closely and to try to link them with exposures in the landslip. Upper Greensand, Cenomanian Limestone and Middle Chalk are all present.

The most southerly block now is of Cenomanian Limestone—the erosion surfaces separating the various divisions

FIG 69

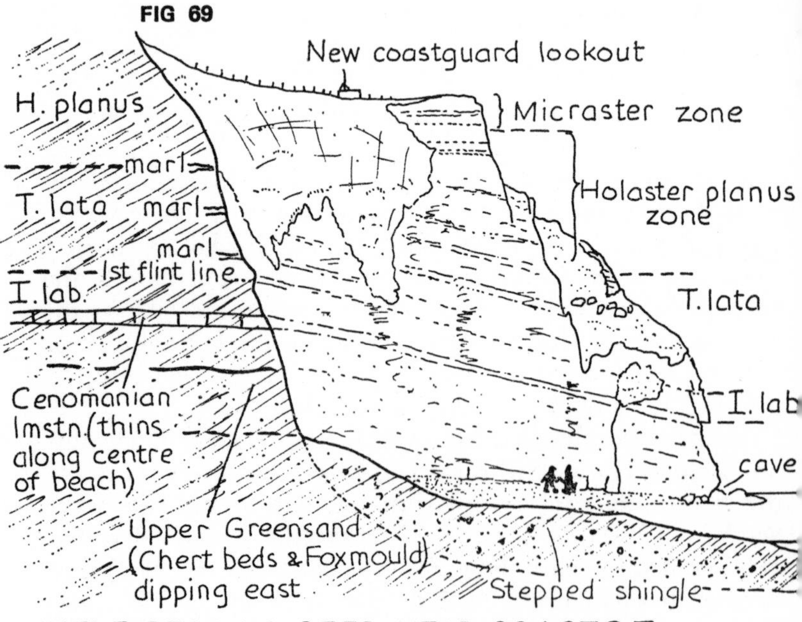

LITTLE BEACH & BEER HEAD 224878 E

can be picked out since each coarsens upwards ending in a stony conglomerate of sandstone and limestone. Reaching 20 feet, the section has obviously fallen from near the Beer Stone adit level.

Beer Head marks the far end of the shingle beach. Here there is a small cave with large coral-like bryozoa (*Ceriopora ramulosa*) forming thick clusters in its roof. If the explorer keeps a close check on the cliffs as he approaches Beer Head, he will see that the Cenomanian Limestones thin along

FIG 70

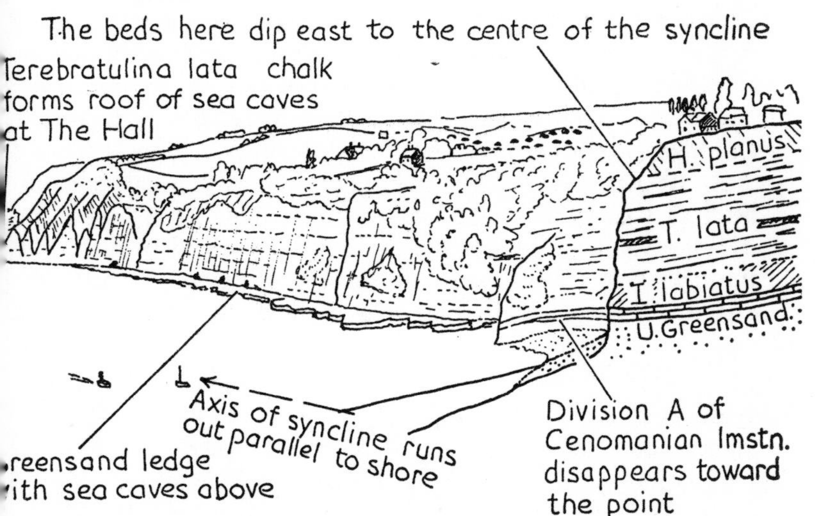

THE CLIFFS WEST OF BEER BEACH 233892 SW

Little Beach (Figure 69) thickening slightly again nearer the head. A similar thinning is found at Beer beach.

BEER BEACH AND WHITECLIFF

Here the thinning can be seen in a small projecting point on the west side of the beach. Working along either side of this feature the explorer can see how Division A thins eastwards and disappears. At the point only Division B separates the Greensand from Middle Chalk above.

Like the Branscombe Mouth Greensand ridge, this is another feature revealing folding movements at the time of deposition. The lower bed Division A was quickly eroded from the crest of a small upfold which occurred then. It is this sort of phenomenon which makes the Beer district so notable geologically.

Looking west from Beer beach there is a prominent ledge at the base of the cliffs. It extends as far as the sea-caves at the Hall and is formed of Greensand. The little sea-caves above it are cut in the Cenomanian Limestones.

Walking eastwards to Seaton, the shore can be followed at low tide, the beds rising up the cliffs until at the east end of Beer beach they form the best section of Middle Chalk in Devon. Beneath them there are more sea-caves, some forming natural arches and allowing you to get through promontories at half-tide, though this is not a practice to recommend! There are compensations if you have to forgo this scramble and take the clifftop path.

Annis' Knob is the chief feature on this path, a conspicuous bluff of white flinty chalk above the village. The base is in Middle Chalk—*Holaster planus* can be obtained from it, but the top 10 feet are Upper Chalk with *Micraster cortestudinarium*. Mid-way up its face is a marked line of flints. Split

FIG 71

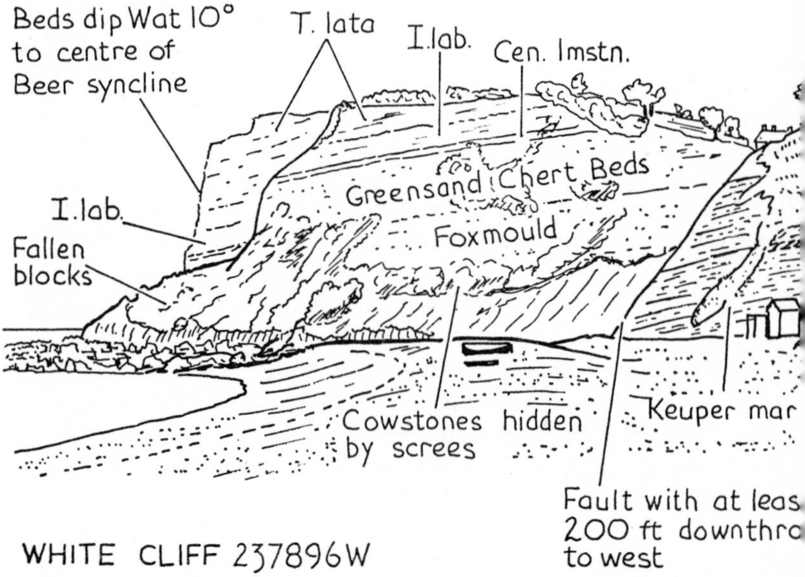

WHITE CLIFF 237896W

open they are almost entirely crust with very little core—unique immature flints whose centres never solidified. Passing over the top of the headland, take the cliff path to the end of Seaton beach for a view of Whitecliff.

By this time the Middle Chalk has risen still further to lie in the top part of the face and beneath it a thick development of Greensand is now visible. The exposure reveals its division into Chert Beds and Foxmould. Together these form 'normal' Greensand in contrast to the exposures at Sidmouth.

The Foxmould is a thick development of yellowish green sands with darker beds at the base, called 'Cowstones'. The green colouring comes from the mineral glauconite, known to form under modern marine conditions near the edge of the continental shelf. The Cowstones account for only 15 feet of nearly 100 feet of Foxmould and their descriptive name comes from the large chalky lumps found in them! The best place to look for these is on the beach, since slips cover most of cliff exposures.

At Seaton Hole a very large slip covers the fault marking the limit of the Cretaceous beds. The movement along it must have been at least 200 feet. In a few yards there is a change from the imposing heights of Whitecliff to a reappearance of red Keuper marls. Standing back on the beach, the walker can pick out two green marl bands running through them, revealing that the soft, crumbly marls are gently folded.

At the far end of the beach the mouth of the river Axe has been pushed eastwards by shingle movement. The history of Seaton and Axmouth is punctuated by attempts to overcome this nuisance which caused an early decline in the once valuable fishing interests of Axmouth.

A more unusual industry here today is pebble-picking. About ten million are picked and sorted by hand each year, and the work is also done at Branscombe. Farmers, housewives and fishermen help to gather the harvest of flint cobbles and a Seaton company sends them all over the world for use in grinding down paints, chemicals, soap powders and other materials.

DOWLANDS CLIFF

The coast path can be followed eastwards across Axecliff

golfcourse to Bindon and Dowlands landslips and through them to Lyme Regis. The right of way was established in a legal battle of 1841-3 when the owner of Pinhay Cliffs tried to prevent access. There are five principal slips here—Ware and Pinhay are probably pre-sixteenth century, but Bindon, Dowlands and Whitlands are nineteenth century.

The later slips were well-recorded because of their interest to the early tourist trade of Lyme and neighbouring towns, and because providence played a hand. William Conybeare, the vicar of Axminster, was a geologist, and Dean Buckland, whose cave research has already been mentioned, happened to be staying at Lyme for Christmas 1839 when the great falls occurred.

The wealth of observations and prints was accompanied by a good deal of writing which favoured less rational explanations than the geological one. Some saw the fulfilment of the Book of Revelation; perhaps not for the first time geologists were branded as 'infidels'! For a while, alarm reduced the tourist trade at Lyme while stories of earthquakes and volcanoes did not improve matters. The natural combustion often observed locally in Jurassic pyrite material was probably the source of some rumours. The landslips were in fact relatively silent and smooth, but not without previous warning.

On Christmas Eve 1839 farm labourers and their families returning home to cottages below the cliff from Dowlands Farm found that the path had begun to sink. By 4.00 am their cottage doors were becoming stuck and the beams settling. The gardens were fissured. During the short time it took to load a farm waggon brought down for their furniture, they found they had to remake the track before they could go up again.

The main slip did not occur until Christmas night itself when some coastguards going on duty stumbled over new ridges in their path, jamming their legs in narrow fissures. Fortunately it was a moonlit night and when they saw the land cracking and opening on every side they soon got back to safety. Down on the beach the scene was equally alarming —the movement disturbing the water, noises of falling blocks, and the appearance of an off-shore ridge similar to the one described at Hooken.

FIG 72

FOUNDERING INTO THE FOXMOULD AT DOWLANDS AND BINDON LANDSLIPS

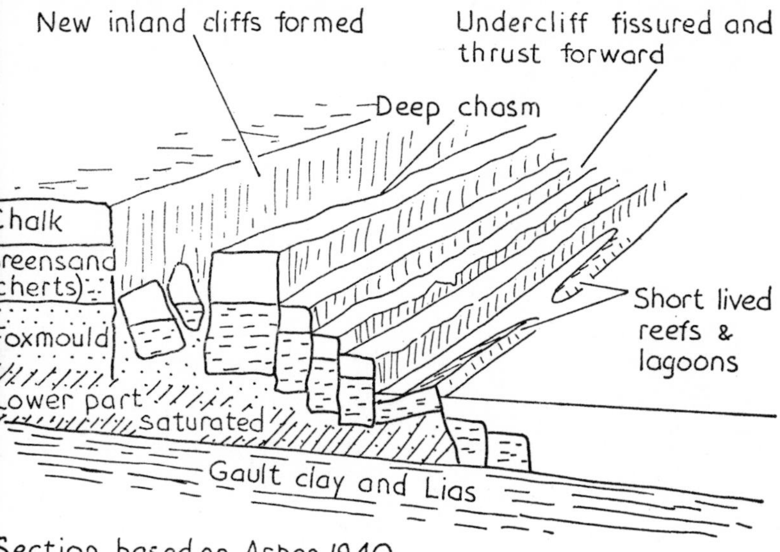

Section based on Arber 1940

The cause was also identical—excessive water making the marl surface below into a slide. The difference from Hooken was the size of the inland chasm here. Geologists who studied Dowlands shortly afterwards suggested this was due to foundering in the Foxmould beds (Figure 72). Their lower area was so saturated that the upper parts subsided into it.

Many features shown in the early prints have, of course, now disappeared. The path through Bindon and Dowlands is densely wooded, the bush growth helping to obscure the remaining pinnacles. But the inland chasms and the new cliff faces are still well-defined and fresh-looking. These slips alone involved twenty acres of land and an estimated eight million tons of rock, while about fifteen acres of fields remained intact on the fallen ground. The Greensand ridge to seawards was soon destroyed by the waves, though some believed it would make a strong outer arm for a new harbour—better than the Cobb at Lyme Regis. The landslips certainly boosted the local trade in other ways however, for when the initial fears were over tourists came in their thousands.

A glance back to Figure 60 will remind us that it is on this part of the coast that the rocks beneath the Cretaceous unconformity change from red Triassic marls to Rhaetic and then Jurassic beds. The explorer should turn off the path through the landslips and make for Culverhole Point where the changes begin. The red marls, over 1,000 feet thick and such an important feature of the coast since Budleigh Salter-

FIG 73

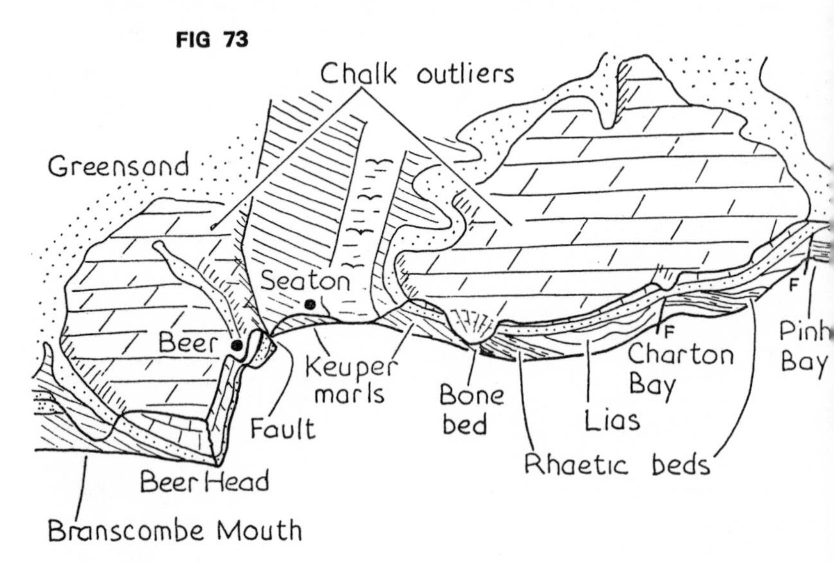

Clay-with-Flints which covers much of the Chalk outcr[op] and disturbances of coastal landslips omitted

CHALK OUTLIERS OF THE EAST DEVON COAST

ton, change in the last 45 feet of the formation to Tea Green marls. These greenish-grey beds are easily seen looking west from Culverhole Point and their gently dipping upper surface reaches the shore in Culverhole Gully.

On the east side of the gully Rhaetic beds appear, formed by marine invasion of the old arid landscape. They are known as 'passage' beds, marking a change from Triassic lands to shallow seas at first and then into the deep-water marine world of the Jurassic. The important thing to remember about them is that they were laid down in the same plane as the Triassic marls. There is no unconformity as with the

Cretaceous covering and Culverhole Gully is an ideal spot to appreciate this point.

The Rhaetic beds begin with a famous bone bed on the east side of the gully. Although it is only two inches thick, within its gritty shales small bones, fish teeth and scales can be found. There are also marine reptile teeth and their fossilised droppings—known as coprolites. The only problem about this bed is that it is sometimes covered by shingle.

East of Culverhole Point are the Lias Beds—the White Lias belonging to Rhaetic times and the Blue Lias to the Jurassic. Old quarrymen referred to those well-bedded rocks as 'layers' and allegedly the name became corrupted to lias—by courtesy of the Dorset accent no doubt! The term was adopted by the pioneer geologist William Smith in 1796. Mainly limestones and marls, they were formed in conditions similar to the Bahaman banks today, calcareous, muddy lagoons. Both White and Blue Lias appear in cliff sections at Wharton and Pinhay Bays but disappear below sea level before reaching Lyme Regis.

FIG 74

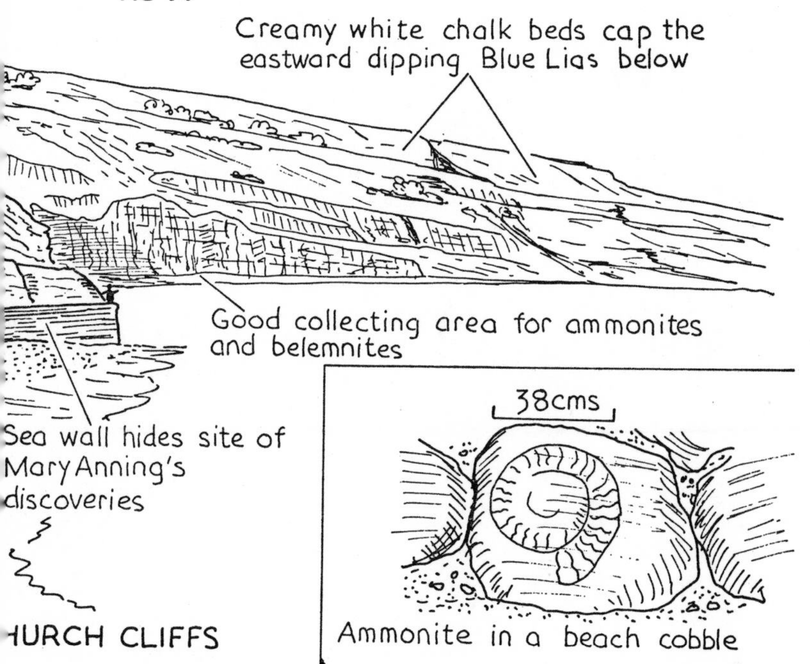

The walker should make for the east side of Lyme Regis where the Blue Lias comes up again in a fold, forming the famous Church Cliffs. Despite the hazards of falling refuse from the town dump, there are worthwhile fossils to be found along the shore. Belemnites and ammonites are common. The cliffs are also famed for the discoveries made there from 1811 onwards by Mary Anning, a local girl. She uncovered skeletons of the vertebrate creatures *Pleisaurus* and *Ichthyosaurus*. The latter were fish lizards resembling dolphins, reaching 17 feet in length and nearly 5 feet in height. Unfortunately, the particular spot where they were found is no longer worked for quarrying and is now partly concealed by a sea wall.

The Jurassic rocks at Lyme were formed in deep water, very murky and poisonously sulphurous, and this led to a lot of pyrite formation. Sometimes ammonites can be found cast in this brass-coloured metal and at others pyrite can be seen burning spontaneously in the cliffs—not to be confused with the refuse burning on the waste tip above.

Eastwards the Cretaceous unconformity can be traced into Dorset, the creamy chalk beds contrasting sharply with the grey Jurassic beds below them in Golden Cap, Stonebarrow and Black Ven.

Chapter 11

The Haldon Hills

Viewed from the air, the Haldon Hills are one of the wonders of South Devon. Their long, flat surface is hardly less impressive from the ground and with their extensive forests and lack of settlement they form a unique district. The Haldons are capped with flint gravels, reason enough for forestry rather than farming. Only the main roads west to Plymouth and Tor Bay, and a few minor routes, interrupt the close stands of timber.

Road materials were the main use of the gravels years ago—carts came up from the surrounding district, Chudleigh and other places, and scores of small pits existed. There were sand quarries, too, working the Greensand lying below the gravels. Nearly all the old quarries are overgrown now and for the geologist hardly any clear faces remain. Even if they are not planted over, quarries in such loose materials soon deteriorate through hill wash.

The basement of the Haldon Hills is the Permian Breccia, already seen in the coastal exposures. Under the northern end they contain Carboniferous fragments, cherts and sandstones, and at the southern end Devonian limestone pebbles. Seismic surveys reveal that the surface of these New Red beds below Haldon is gently folded in a series of E-W ridges, and that the whole feature also has a general westerly dip.

SCHOOL WOOD QUARRY

The most distinctive feature of the New Red beds at Haldon is the Permian lava which occurs all along the eastern slopes. The most accessible exposure is in School Wood quarry near Webberton Cross, Dunchideock, where the reddish coloured lavas can be seen dipping E at 10-15°. They were erupted into an area of wet water-laid sands which now appear as veins filling the cooling cracks in the lava. In the finer grained lava here small red crystals of the mineral iddesite can be found. In the higher parts of the quarry the lava is riddled with gas cavities and contains a lot of chlorite and calcite.

Pebbles of similar nature appear in the sandstones at Shaldon's Ness beach, suggesting they are the same age. This poses an interesting problem because the Haldon lavas were formed at the beginning of Permian times. If the Ness beds are early Permian, too, then the earlier New Red material south of Shaldon must be Late Carboniferous in age. The arid basin was in being, and filling up, well before Permian times began.

School Wood quarry was used a great deal in the inter-war years but now only occasionally. Its warm-coloured rocks are contemporaries of the Rougemont lavas in Exeter. The widespread demand for volcanic stone in Exeter buildings was generally met by Rougemont and other sites nearer than School Wood.

THE HALDON GREENSAND

Resting on the New Red Sandstone beds, the base of the Haldon Greensand is another part of the Cretaceous unconformity (see Page 134). The Greensand varies in thickness, from 16 to 84 metres, often exceeding the thickness found in East Devon. Both Greensands are of the same age and since Haldon was nearer the coast in those days it ought in theory to be the thinner of the two. Hollows in the uneven base are probably the cause of this anomaly.

Fossil molluscs in the Haldon Greensand are thick-shelled, evidence that they lived in disturbed water near the shore. The beds also contain compound corals, pebbly and granitic material. Granitic pieces could only have come from Dartmoor, an island in those days, and they prove that the Haldon Greensand was formed less than 8,000 metres from the coast.

The Greensand is easily picked out around the hillsides, making a marked change of slope with the New Red beds below. The steeper Greensand cover is gradually being

FIG 75

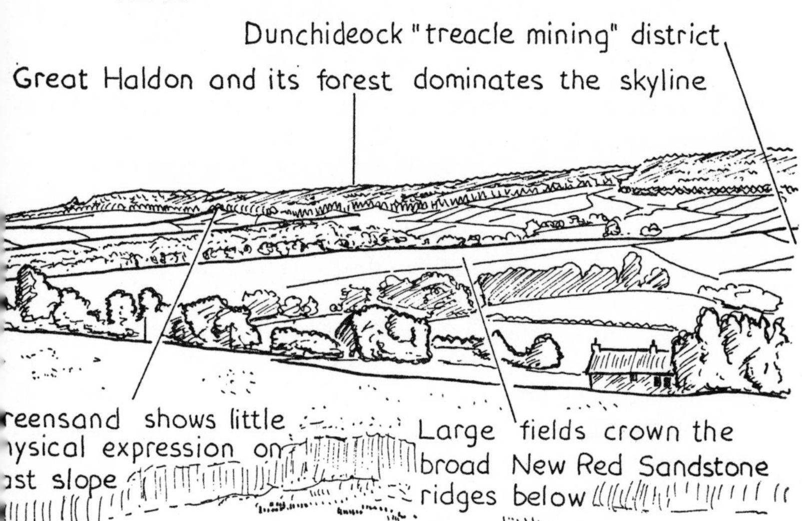

stripped away but it remains least damaged on the north-east slopes since the gently dipping Permian surface sends ground-water out on the opposite side of the ridge. There, facing south-west, the base of the Greensand is blurred and it is often found washed down onto the Permian slopes.

Now rather deteriorated, one of the best and thickest exposures of Greensand used to be the sand quarry at the head of Smallacombe Goyle on the east side of Little Haldon. The only reasonable site today is in Buller's Hill quarry, described below.

THE HALDON GRAVELS

Although Dartmoor was an island in Greensand times, it was of course soon submerged by the Upper Chalk sea. It left a thick flint-filled cover of chalk on Haldon (Senonian age, Page 145). Fossils found in the Haldon flints today show the covering zones were nearly 600 feet thick.

The story of how this chalk disappeared and the later history of the flints it left behind explains why three different gravels can be recognised on Haldon today :

Surface gravels disturbed and washed in the Ice Age.
Gravels moved by a river; containing worn flints and sandy beds.
Undisturbed gravels with fresh flints.

When land conditions returned in Eocene times there was intense tropical weathering—the chalk was dissolved away leaving a massive cover of flint gravels. The bottom part of this mass was never disturbed again and forms Haldon's older gravel now. The upper part however was reworked by a river, which added its own tourmaline pebbles as proof that it came from Dartmoor.

Kaolin (china clay) minerals were also brought down by this river. Some of them sank into the gravels, even as far as the undisturbed beds below, but most of the clay went well beyond Haldon to form the ball-clay deposits of Wareham in Dorset.

FIG 76

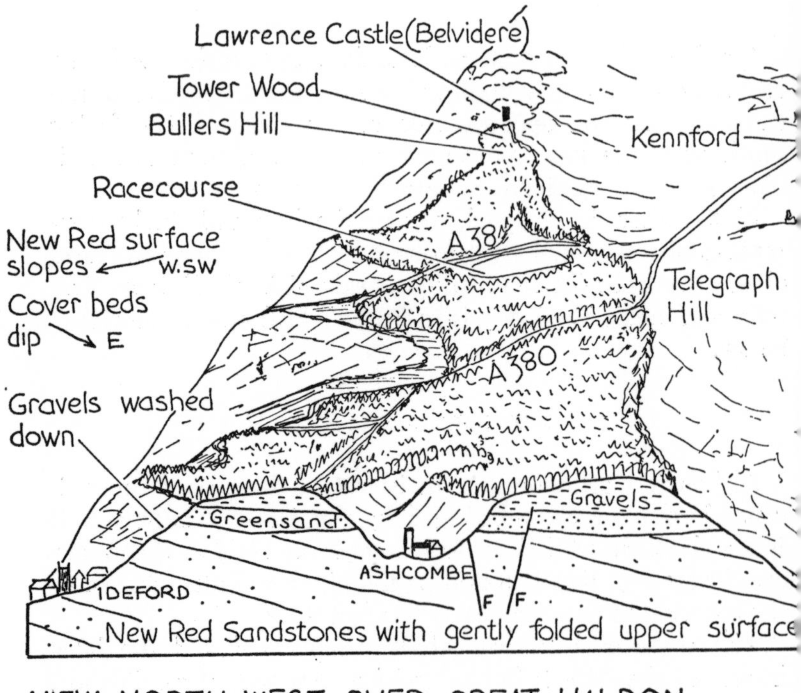

VIEW NORTH WEST OVER GREAT HALDON

The river flowed over Haldon until Oligocene times, when the Sticklepath fault and the formation of the Bovey Basin separated Haldon from Dartmoor. The Bovey Basin then became the new site for ball-clay formation (Chapter 12).

The final episode in the history of Haldon's now isolated gravels was the usual slumping and shattering of the Ice Age. Flints washed off the margins at that time now cover the Greensand and Permian slopes below, blurring the surface distinction between the various gravels and making their interpretation very difficult.

TOWER WOOD AND BULLERS HILL

Tower Wood quarry lies south of Haldon Belvedere, on the east side of the road (876857). Worked for gravel, its flints belong to the residue left when the chalk was dissolved away. These older gravels vary—their greatest thickness on the hills is 60 feet and at Haldon racecourse they reach 30 feet. Quite a variety of fossils can be found in them, including sea urchins.

To the SSE there is a larger quarry at Bullers Hill. Cars can be parked at the junction with the old Chudleigh-Exeter road and the quarry lies west of the Forestry Depot at 886847. The section here is more extensive, including Greensand, undisturbed and re-worked gravels.

Bullers Hill quarry shows that the base of the gravels is level—there is no cambering of the beds over the Haldons ridge. Large blocks of Greensand litter the lower side of the working, revealing later silica deposits in their rounded lumps of chalcedony, and a rich fossil life. *Pecten* is abundant and there are good examples of worm burrows.

A reasonable cross-section of Haldon is provided by the A38. On the east side the New Red Sandstone beds are seen near the old toll-house. Up the hill there are more exposures of them until, on the inside of the last bend before the summit, the road passes up into yellowish sands and gravels. From the dual carriageway down the western side the spread of the flints downhill can often be seen in ploughed land north of the road. Another feature of the western side is the belt of Devonian limestone lying beneath the Permian. It crosses the A38 at the lower end of the dual carriageways.

FIG 77

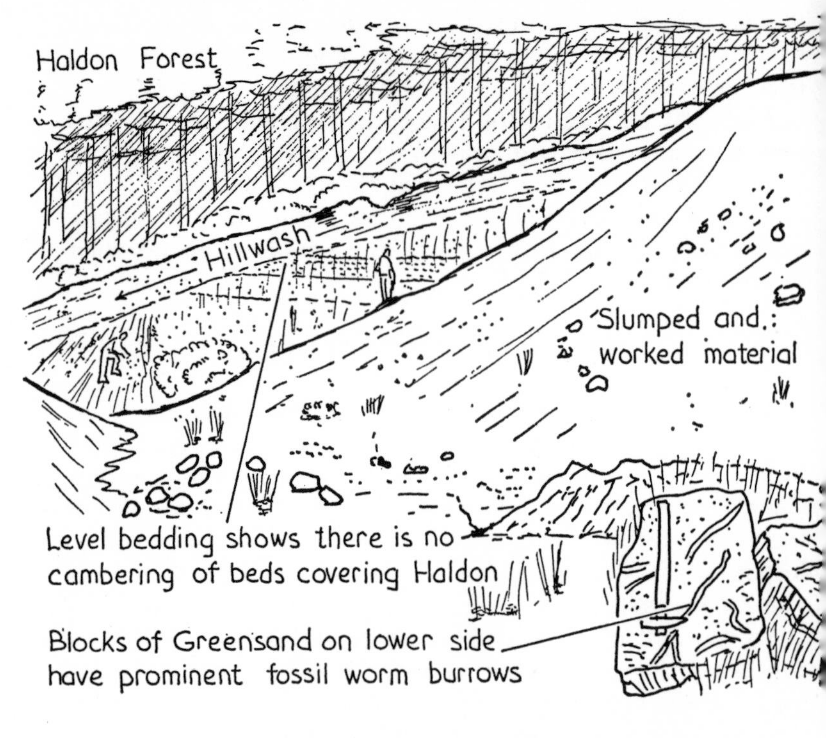

BULLERS HILL QUARRY 883847 N

Large blocks of limestone stand on the verges and old lime-kilns in the central reservation have been separated from their quarry.

TREACLE MINES

It was on the east side of the Haldons that this perennial legend came to life again in 1969—School Wood quarry was implicated in the story which appeared in *Devon and Cornwall Notes and Queries* (Vol 31, part 3). Dunchideock treacle mines were described, their treacle coming from beds rich in the remains of moss-like plants which lived millions of years ago. The sources were said to be beneath Dunchideock House and in quarries above the village. The beds were sought at night, being detected by their fluorescence in ultra-violet light. But Dunchideock House stands on New Red beds (moss-like plants in an arid desert?) and the quarries, which must include School Wood, are in Permian lavas! The legend has no geological foundation.

Nevertheless, it is a peculiarly widespread story, occurring all over the country, and other alleged sites in South Devon include Daccombe, near Newton Abbot, and Tamerton Foliot at Plymouth.

Chapter 12

The Bovey Basin

By a curious coincidence, the two extensive flat districts of South Devon lie within sight of each other—the Bovey Basin and the Haldon Hills. The Bovey Basin shares the youthful geology of the Haldons, their gravels and sands, but differs from them in its thick clay deposits and its low elevation. Its story is not one of a capping laid smoothly across the rocks below. Rather is it a history of tearing and breaking, subsidence of the land and the white waters of a great lake, for the Bovey Basin lies along the great Sticklepath fault.

THE STICKLEPATH FAULT

One evening in 1955, Sticklepath, near Okehampton, was shaken by an earth tremor. The village stands directly on the fault, a great crack which runs right across Devon, and the most famous of a series of NW-SE tear faults dating from mid-Tertiary times.

Faults are lines of weakness and fracture, rarely formed in a single movement. Minor adjustments can spread over millions of years, as the 1955 tremor at Sticklepath shows. Stresses build up gradually in the rocks and their typical sudden release causes the earth tremor. Most faults produce vertical movements but tear faults are the less common horizontal type.

Scissors and paper can easily reproduce this sort of thing. Cut out an outline of Devon and mark on it the positions of Sticklepath and the river Bovey. Draw a line through them, right across the piece of paper. Now cut your outline in two along it and you have a Sticklepath fault. Place the two pieces of paper side by side on a flat surface and slip them to and fro horizontally along the junction. You are making the

MAP 12

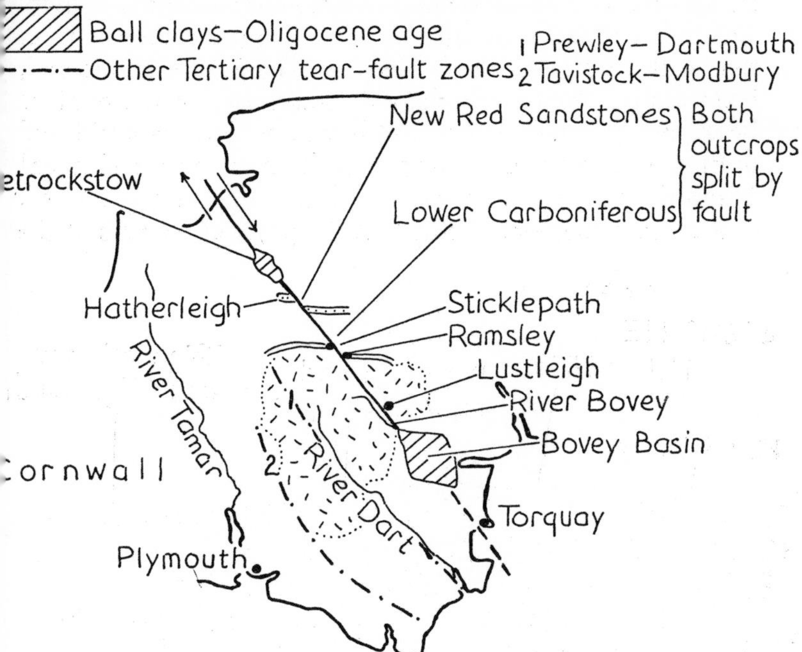

movement of a tear fault.

The Sticklepath fault probably begins in Tor Bay, off Berry Head, and passes up through the Torre Valley and Aller Vale towards Newton Abbot. Since the movement along it was about $1\frac{1}{4}$ miles, the Devonian outcrops of Kingskerswell and Torquay could have been separated by it. Try drawing them in on your original map and then, moving the pieces, link the two outcrops up again.

The Bovey Basin was certainly the work of the fault activity. The break can be traced up the river Bovey at Lustleigh, crossing the Teign just west of Chagford, and then passing near Okehampton and Hatherleigh it continues out to the coast west of Bideford.

THE ROCKS OF THE BOVEY BASIN

Diamond-shaped, its twenty-eight square miles drained by

the rivers Teign and Bovey, the basin is a focal point of South Devon geology. Around it lie slates, limestones, shales, sandstones and granite—Devonian, Carboniferous and Permian. The local villages were sited on these margins rather than on the heath itself. Chudleigh Knighton is a good example and, incidentally, stands near the only small area of the original heathland left today. The heath was a place to avoid and until the nineteenth century brought road, rail and canal improvements, use of the clays within the basin was limited.

It is from the borderlands that the physical features of the

FIG 78

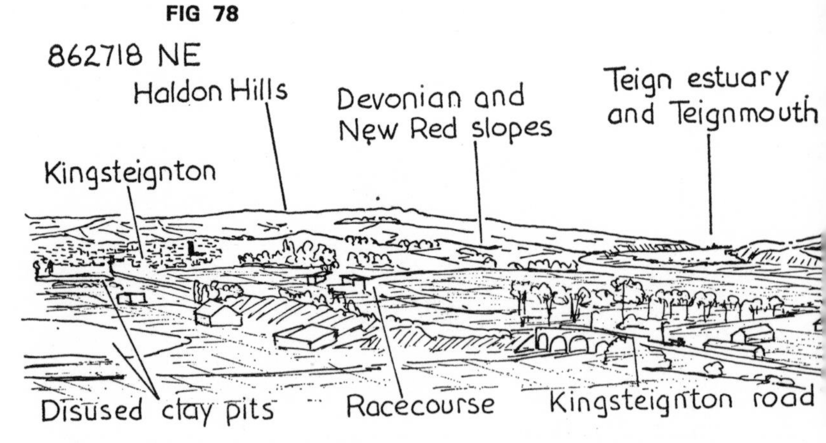

PART OF BOVEY BASIN FROM KNOWLES HILL

basin should first be studied and the road around Knowles Hill, Newton Abbot, is a good vantage point (Figure 78).

From Knowles Hill make for the east side of the basin at Kingsteignton. There various quarries, sand and gravel pits provide exposures of the same Greensands and flint gravels described on Haldon. With the clays these make the three formations of the basin :

The Bovey Beds : clays with lignites and sands : Oligocene Age.

The Aller Gravels : flints and sands : Eocene Age.

The Greensand : yellowish-green in colour : Mid-Cretaceous Age here.

All three can be seen on the east side of the basin since there is no faulting there. Their outcrops, continuing west of

FIG 79

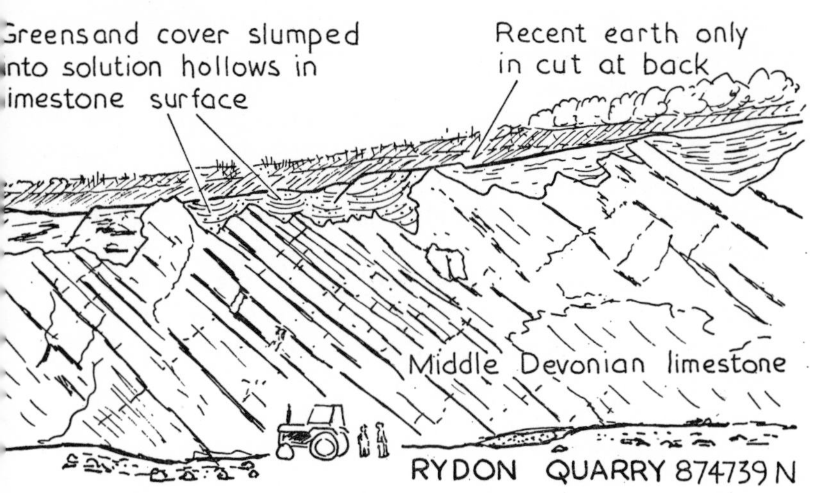

Haldon, must have been bent over into the Bovey trough as it subsided.

At Rydon quarry, 874739, the Greensand lies in pockets across the top of Devonian limestones. Down-folds visible in the sands show how solution of the limestone below encouraged them to slump into little hollows all along the top of the quarry. The limestones were the product worked here and the quarry is interesting because its two sections are connected by a tunnel.

BABCOMBE COPSE AND SANDS COPSE

North of Kingsteignton the Greensand itself is worked for building purposes at Babcombe Copse. The quarry, owned by Kingston Minerals, lies in woodland west of the first RAC telephone along the A380. Sand quarries are often unstable so permission is needed to enter them and they cannot be worked too close to a roadway.

The Babcombe Copse Greensand was formed even nearer the shore than Haldon's. Twenty-five feet thick, the beds are pebbly and gravelly with chert beds containing silicified *Orbitalina* and *Exogyra.* A small trowel is a handy weapon here to scrape moss off the face and reveal the true sand colours.

The Greensand dips westwards at 5° towards the Bovey Basin. In a complete section it would eventually pass below the Aller Gravels, but Babcombe Copse is an isolated pocket. However, returning to Kingsteignton, the gravels can be seen in another quarry at Sands Copse, 866755. These, too, dip westwards and beyond the lane vanish beneath the thick Bovey clays, the main filling of the basin. As a rough guide the slope of the hill at Sands Copse is about the same as the westerly dip.

The Sands Copse gravels are part of the Aller Gravels, named from their more extensive outcrop towards Kingskerswell, and the equivalent of the re-worked gravels of Haldon

FIG 80

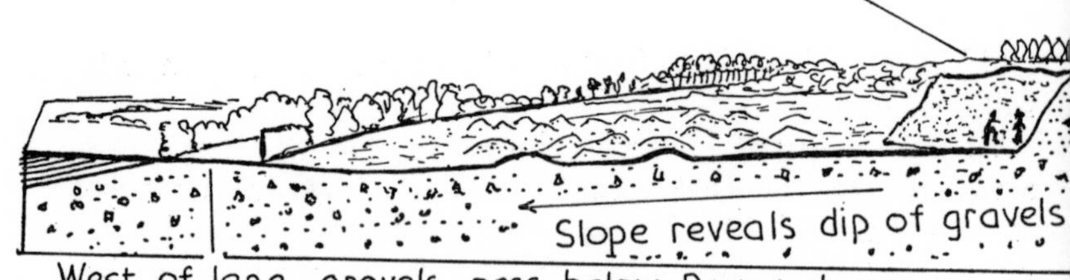

SANDS COPSE GRAVEL PIT 867755 N

(see Table of Strata). Before studying the important clay industry, the walker should visit the gravel workings there in the Royal Aller Vale and adjacent quarries. He should also look at Chudleigh Knighton church—it seems so out of place in Devon, its walls of split flint cobbles recalling south-east England in style.

ROYAL ALLER VALE QUARRY

The Greensand outcrops on the hillside above the quarry and, dipping west beneath it, is just exposed in the bottom of the working. The impressive feature of the quarry is the mixed nature of the Aller Gravels. There is no hope of selective working here so the material must be separated by screening and washing. Channelling and other features in the beds

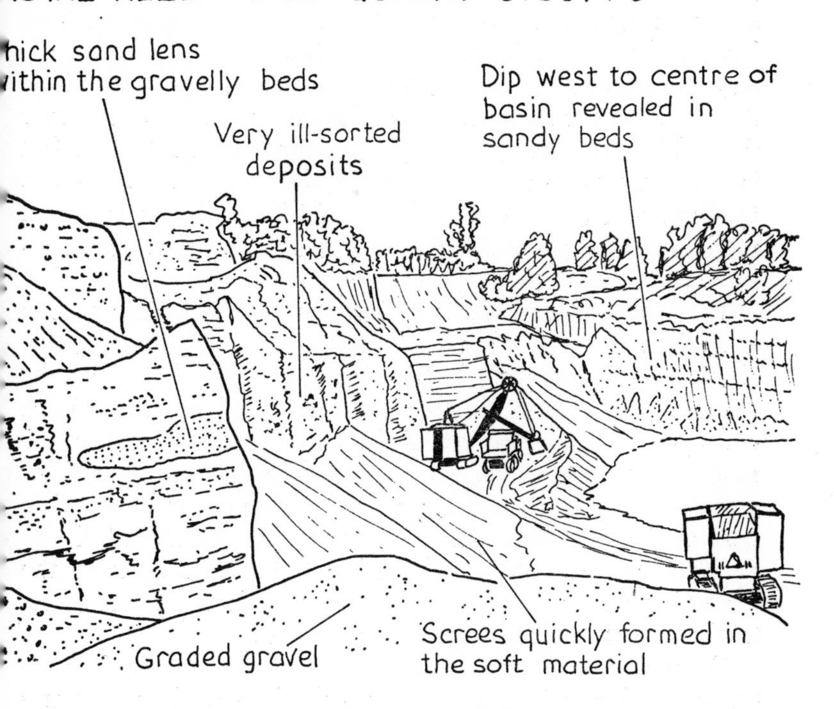

reveal that a river landscape existed here in Eocene times, its braided channels wandering over a flattish plain. In the hot Eocene climate with its heavy downpours, a very ill-sorted deposit was produced. Flints, greensand, cherts and quartz make up the gravels.

THE BOVEY LAKE

The Bovey clays are the most important deposits of the area. They also contain sands and lignites. The clays accumulated in a great depression formed by faults. The Sticklepath fault passes down the centre of the basin and others border the western and southern margins. Subsidence began following the river plain of Aller Gravel times and the basin soon turned into a lake, its waters filled with brown and white clay minerals. The waters were so acid that little life could survive in them, so the beds contain no fossils.

Floods washed the trunks of sequoia trees and various plant remains out into the waters where they became saturated and sank to the bottom. These organic remains formed the lignite beds, black bands of waste as far as the clay industry is concerned, but once mined a mile south of Bovey Tracey for house fires. There, in the 1860s, over thirty beds of lignite, ranging from three inches to four feet in thickness, were recorded in a 125-foot section. Lignite was used again as an emergency household fuel between 1939 and 1945 but only patriotism could put up with its sulphurous smell!

The water of the lake is still a source of argument among geologists. It could have come from altered granites, the china clay deposits of Dartmoor simply washed here by the rivers Bovey and Teign, but neither valley contained large areas with this special type of decomposition. Could it be the result of weathering on the Dartmoor surface in general, or perhaps of the Culm slates near the margin of the basin? One current idea is that the lower beds are weathered Culm material but that the chief source of the clays remains the general weathering of the Dartmoor surface. There is certainly a great deal of kaolinite in the basin.

Compared to Dartmoor's china clays, the Bovey deposits are much less gritty. There is no need to wash out quartz sand and mica from the worked horizons since this was done naturally while the clay was being carried here by the rivers. The clays are generally white and brown, siliceous, with pink-stained sands a feature of the lower layers. Faults can be found in them; some following the Sticklepath trend prove later movements along that fault in particular.

The name 'ball clay' given to the beds comes from the ancient hand methods of working. The beds were dug in layers, cut into squarish blocks or 'balls', spiked with a pointed stick and heaved on to wagons by hand. Hand-worked pits were worked in levels to control quality, the gangs moving from shelf to shelf by ordinary ladders. Vertical shafts were also used for underground working but these could only rob a limited area around the bottom of the shaft. Gradual movements in the clays often pushed them out of alignment and made haulage difficult.

The earliest working here is unknown but as late as the

seventeenth century production was very small since the ancient methods were limited to trenching and water was a constant problem. It was the knowledge of the Dutch potters of Delft, becoming available in this country when the House of Orange came to the throne, which provided the impetus for an English pottery trade in the eighteenth century. A few

MAP 13

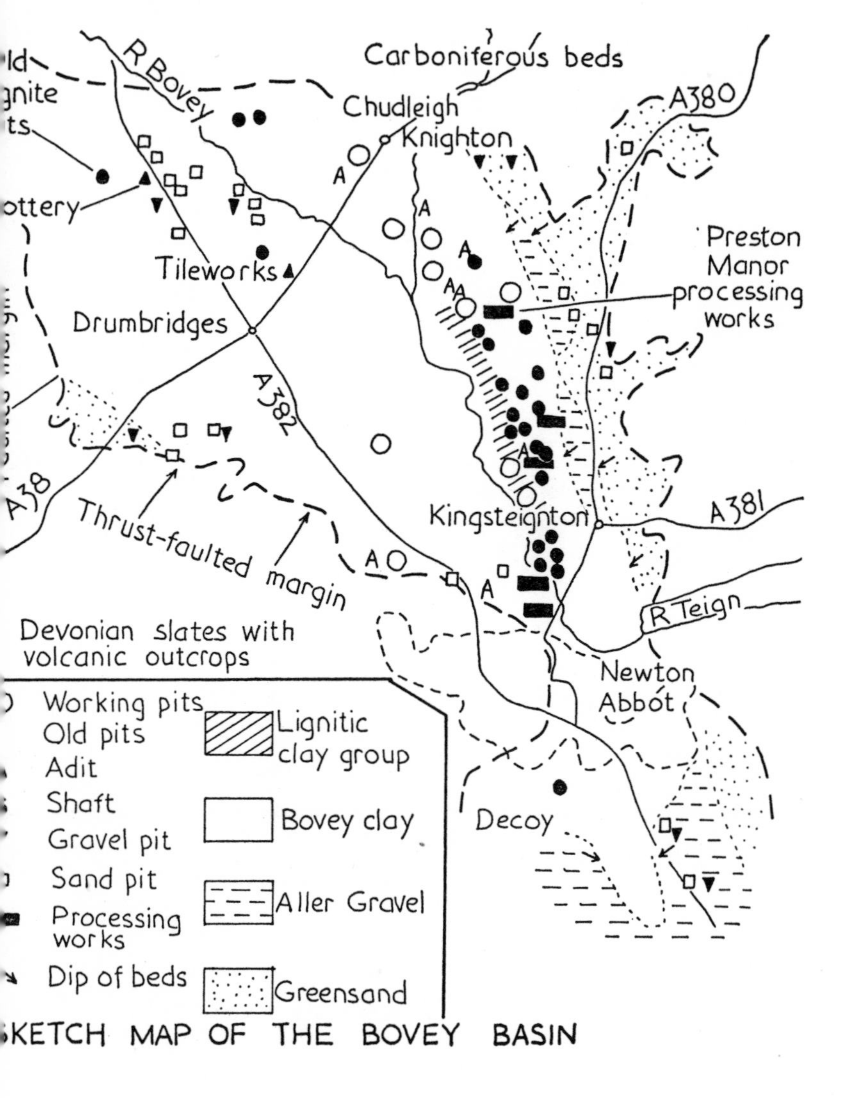

SKETCH MAP OF THE BOVEY BASIN

hundred tons a year of South Devon ball clay were in use by 1740, but in 1785 10,000 tons left Teignmouth in 115 boats. Potteries developed locally at Bovey Tracey and Aller Vale.

Ball clay is highly plastic in nature and, since some of the beds burn white when fired, it finds its largest modern markets in vitreous china sanitary ware and in wall-tile manufacture, followed by tableware and electrical porcelain products. It is marketed in forty-four different qualities, 77 per cent going overseas where Italy is the largest buyer. Today, South Devon provides 70 per cent of the national ball clay production of 725,000 tons (the rest comes from North Devon and Dorset) and Teignmouth exports 350,000 tons a year.

It is rather amusing now to read early estimates of the depth of clays in the basin, only 400 feet, since evidence now points to a basin 3,500 feet deep. To accumulate this quantity

FIG 82

BALL CLA[Y] WORKING

OLD AND MODERN OPENCAST METHOD

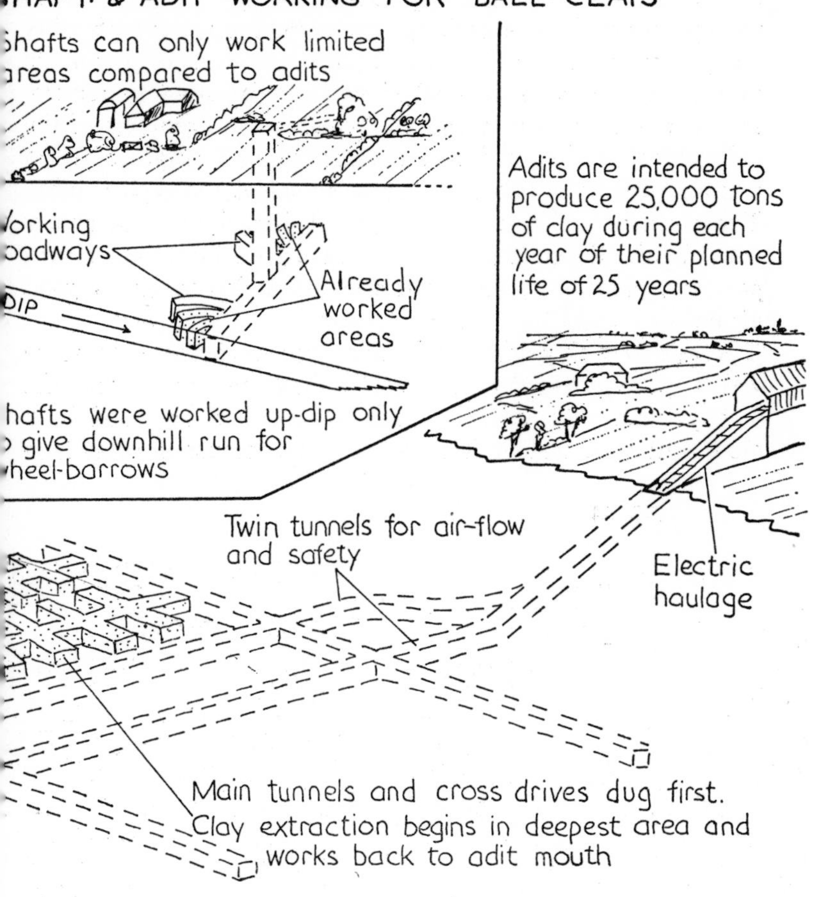

of material, the sediments must have subsided under their own weight as well as along faults associated with the Sticklepath fault. Only a small amount of it is useful clay, of course. At the old Heathfield vitrified clay-pipe works in the centre of the basin, 2,000 feet of poorer gravelly material lies between the A38 and the 400 feet of workable clay below.

PIT AND ADIT WORKING

Modern workings are concentrated down the east side of the basin (Map 13). Clay is obtained from opencast pits and from deep adits driven westwards towards the centre of the basin. East of the Kingsteignton-Chudleigh road the clays were poorer and more working was in open pits, but both adits and pits exist west of the road. Adits allow levels of particular ceramic qualities to be worked without the need to remove all the overburden beforehand. Avoiding the problems of the old vertical shafts, modern adits now reach a length of 2,600 feet and depths of 600 feet. Figure 83 shows the typical plan. The main tunnel is driven steeply at first, then more gently with depth. From it side levels are driven and then return passages, parallel to the main one, are added to give a good air flow.

A visit to a clay adit is a fascinating experience. A separate walkway is provided alongside the haulage tunnel and the first 50 metres of vertical depth is supported by wood in a close timbering system. This is necessary because the clay is so

FIG 84

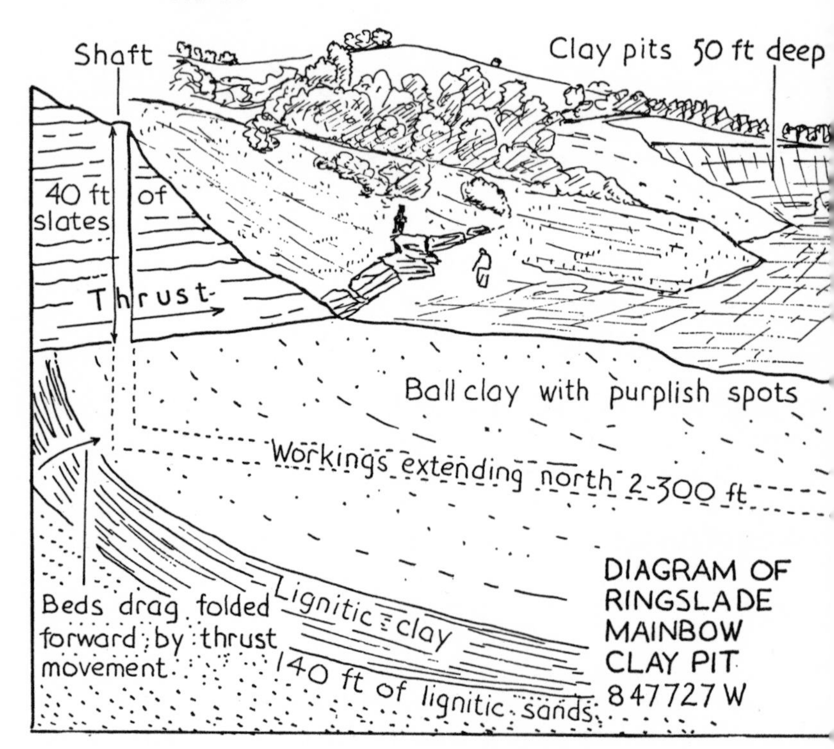

plastic that it would squeeze through like toothpaste! At lower depths steel colliery arches are employed, standing on bottom timbers which stretch the full width of the adit drive and are again backed by wood boards. These measures effectively combat squeeze pressure and permit rapid recovery of the clay from the working areas.

Slips are rare in these adits and in clay materials are usually slow in any case. There is less of the danger associated with solid rocks where pressure builds up undetected until its sudden release.

RINGSLADE-MAINBOW PIT

A mass of Upper Devonian slate outcrops westwards from Newton Abbot, its curving north and south boundaries marked by thrust faults. The outcrop is broken by exposures of volcanic rocks which give the countryside here a distinctive appearance. Numerous small hills lie in strings along the ridge tops, each one marking a volcanic outcrop.

The Ringslade-Mainbow claypit lies below the boundary of the slates near Highweek. Figure 84 shows how the clays here were worked both by opencast and shaft methods. But the shafts lay up the hillside from the pit and had to sink through 40 feet of slates to reach the younger clays below.

This topsy-turvy situation is the work of the thrust fault. The slates bordering the basin were carried northwards overriding the younger clays and partly burying them. The old clay-miners who discovered this secret were certainly remarkable men! Their work at Ringslade-Mainbow shows that the Tertiary fault which formed the Bovey Basin continued to move after the clays had accumulated in it.

The Ringslade-Mainbow clays are also noted for their coloured spots. Since the colours are similar to those of the slates, the spots may be pieces of slate broken off and ground into the clays during the thrust movement.

THE CHIPLEY PILLOW LAVAS

The most interesting volcanic outcrop in the slates is at Chipley. Turn northwards off the A383 at Chipley and walk up to the next junction in the lane. Turn left and, with

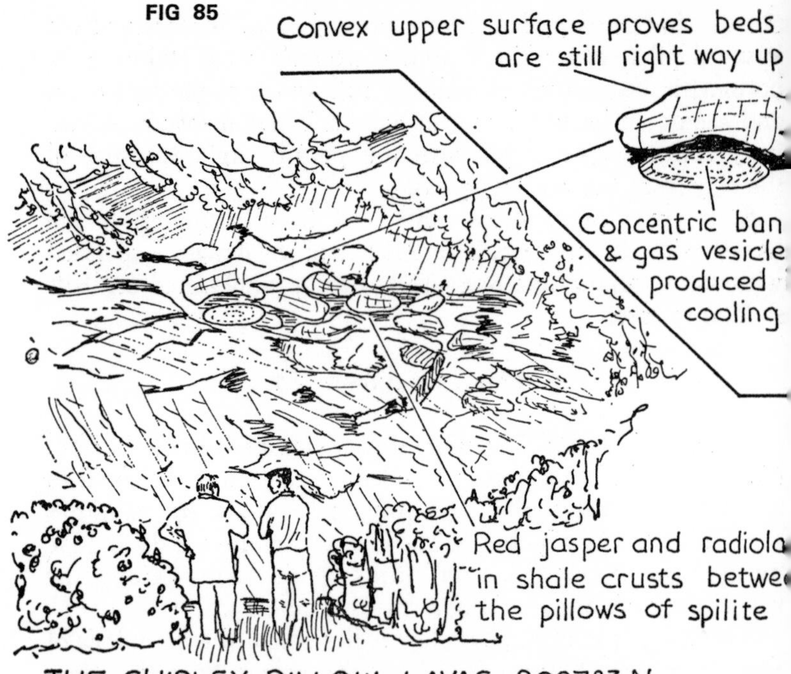

FIG 85

THE CHIPLEY PILLOW LAVAS 809723 N

the farmer's permission, enter the first field gate on the right. At the top of the field a small quarry reveals the best pillow lava exposure in South Devon. Alternatively there is another quarry right by the road, a little farther north along the same lane.

Pillow lava, as we have previously seen, is formed by an eruption under the sea. The lava breaks into lumps instead of flowing in a sheet, and these lumps, falling to the sea bed, accumulate in curving masses like a heap of pillows. Each one forms with a convex upper surface and since the Chipley pillows still have this on top it proves they have not been overturned since Devonian times.

And what tough material the pillows are. Obtaining a sample of the spilites they are made of is very hard work. Spilites are a form of basalt, grey coloured with white spots. Each spilite pillow has a slate or shale crust and thin films of red jasper occur in the latter. Concentric lines of holes (vesicles) can be seen where gases escaped before the lumps became solid.

BICKINGTON AND CHUDLEIGH

Everyone who travels the A38 will be familiar with the uphill pull through Bickington. The road then follows the contours eastwards, looping around the hillside before it descends again into the Bovey Basin. This contoured road lies very near another part of the thrust seen in the Kate Brook at Chudleigh (below). Here it carried the slate and volcanic beds of Telegraph and Ingsdon Hills over the Carboniferous beds which

FIG 86

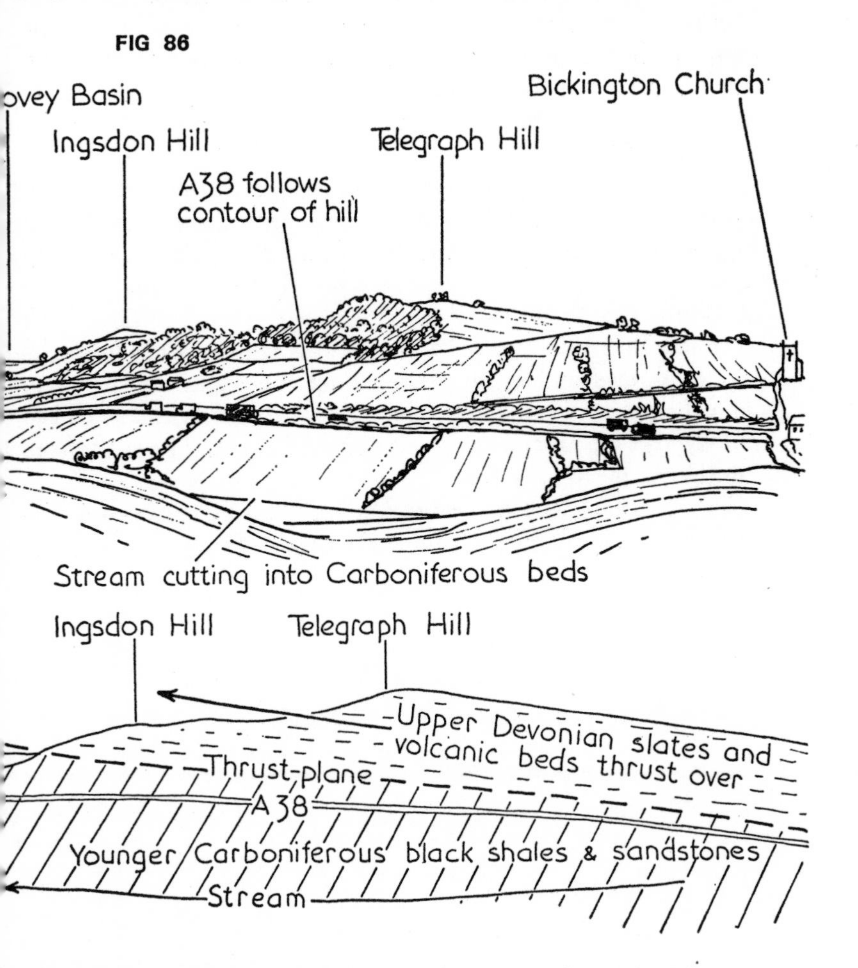

THE BICKINGTON THRUST PLANE 795739 SE

form the valley below the roadway (Figure 86). Rolling green fields mask all surface evidence of this zone of movement and earth tremor now—another example of how well-hidden the features of South Devon geology can be.

The thrusts are an important element of the area, however, and their work can be seen again on the east side of the basin at Chudleigh Rocks. The Chudleigh limestones are well-known for their quarries, caves and rock climbing. The thrust passes beneath them and can be seen near Palace quarry, which itself provides the usual rich, coral life of these old Middle Devonian reefs. Following the stream below the quarry, the explorer soon finds the beds change to green slates and, since these are the Upper Devonian Kate Brook beds, once again the thrusts have carried older rocks over the top of younger ones.

The Kate Brook cuts a beautiful gorge through the limestone area. It must have lowered itself into these beds from a higher level, helped by solution and the collapse of any caverns in its path. Follow the track above it to the summit of Chudleigh Rocks where there is an extensive view back over the Bovey Basin. Many clayworks can be seen, with the tall chimneys at Heathfield marking the centre of the area. The thrust on the opposite side at Bickington is easily picked out by the dip in the skyline to one side of it. There the A38 climbs out on its way west.

Every square foot of the ground in front of you reveals the intimate bonds between man and his environment—a heritage millions of years old.

Glossary

AMMONITES—Coiled-shelled molluscs which existed from Triassic to late Cretaceous times. They were active, evolved rapidly and make excellent zone fossils for the Mesozoic deposits.

ANTICLINE—Upfolded beds, arch-like with the older layers inside the younger beds.

AQUIFER—Porous bed capable of acting as a reservoir of ground water for water supply, eg, a sandstone.

BELEMNITES—Squid-like molluscs with internal calcareous shells. Their cigar-shaped guards are common in Jurassic and Cretaceous beds.

BRACHIOPODS—Marine shell-dwelling creatures. The chief characteristic of the thousands of species that existed and the 200 or so which still do is the symmetry of the shell. Each valve is symmetrical and often ribbed but the two valves are quite different (c.f. **Lamellibranchs** below).

BRECCIA—Coarse angular cemented rocks consisting of fragments derived from older beds or some sort of activity like faulting or crushing.

CAINOZOIC—Literally 'the Recent Life', means the Tertiary era in geological history and includes Eocene, Oligocene, Miocene and Pliocene.

CLEAVAGE—The way minerals split along planes related to their crystal structure or the way certain fine-grained rocks split.

CONGLOMERATES—Large rounded fragments of other rocks cemented into a finer grained mass, eg, the Pebble Beds.

CRINOIDS—Marine animals fixed to the sea bed by a stem; called sea-lilies due to their appearance.

CULM—Name given to rocks of Coal Measure (Carboniferous) age in Devon. The thin bands of coal in them are referred to as Culm in early mining and trading accounts.

DIP-SLOPE—Gentle sloping surface formed by the top of a tilted bed. (*See* Scarp).

DOLERITE—Also known as Greenstone. Hard dark-green rock found in volcanic intrusions. Crystalline appearance.

DRIFT—Superficial deposits covering the solid rocks. Generally applied to clay and boulder deposits of ice-sheets and glaciers.

ESCARPMENT—*See* Scarp.

FELSITE—Igneous rock, reddish-coloured with quartz and felspars in it.

FELSPARS—Aluminium silicates basically; common rock forming minerals and important in granite.

FLINT—Siliceous rocks found in nodules and often acting as fossil casts. Believed to be solidified remains of siliceous sponges.

FRASNIAN—The lower part of Upper Devonian time.

GONIATITES—Coiled-shelled molluscs living from Lower Devonian to Upper Permian times, valuable like the ammonites as zone fossils, but less decorated with frills.

HEAD—Mantle of rock waste washed downhill under cold climatic conditions.

IGNEOUS ROCKS—Rocks which were molten at some time in their history. Termed 'intrusive' if they cooled below ground, 'extrusive' ie, volcanic if they did so at the surface.

INLIER—Area of older rock appearing up through a surrounding of younger rocks.

INTERGLACIAL—Warm climatic phase between two periods of glaciation.

LAMELLIBRANCHS—Bivalved shells like the brachiopods but the two valves are identical, mirror images of each other.

LAMPROPHYRE—Igneous rock with abundance of iron and manganese minerals, altered calcite and felspar products.

MESOZOIC—The 'Middle Life', the era containing the Triassic, Rhaetic, Jurassic and Cretaceous systems.

METAMORPHIC—Rocks originally sediments but transformed by contact with heat, pressure or active, migrating fluids.

MICA—Important mineral with a splendid lustre; splits into thin, elastic plates. White mica is called muscovite, black is biotite.

NAMURIAN—Name given to the lower part of Upper Carboniferous time.

OUTLIER—Detached patch of younger rock lying amidst older beds.

PALAEOZOIC—The era of 'Ancient Life', including Cambrian, Ordovician, Silurian, Devonian, Carboniferous and Permian times.

PLEISTOCENE—The Ice Age part of the Quaternary era of geological time—the era in which we live today.

QUARTZ—A very common mineral; important in granite (Silicon dioxide, SiO_2).

QUARTZITE—Metamorphic rock, a recrystallised sandstone largely formed of quartz; or a sandstone cemented by silica.

RAISED BEACH—Beach formed by a former sea-level and now clear of wave action. Uplift of the land or lowering of the sea could both account for it.

SCARP (ESCARPMENT)—Ridge with steep slope one side (Scarpface) and gentle one the other (Dip-slope), being the end and upper surface respectively of a tilted rock layer. Common in chalk and younger limestone districts.

SCHISTS—Metamorphic rocks with a streaky appearance; wavy, uneven surfaces: they split easily because they contain a good deal of tabular minerals such as mica.

SEDIMENTARY ROCKS—Formed from the accumulated remains of other destroyed rocks and deposited naturally in seas, rivers, lakes.

SOLIFLUCTION (HILLWASH)—Soil creep, generally associated with movement of thawed-out surface layers in cold climates, but the headstones of any sloping churchyard will confirm it occurs in present climates to some extent.

STROMATOPOROID—Large, mound-like remains secreted by animals living in central canals for their protection. Classified by the system of pillars supporting the concentric layers.

SYNCLINE—Down-folded structure in rocks, the lower beds surrounding the upper.

TERTIARY—*See* Cainozoic.

TRILOBITES—Extinct marine arthropods. Important as zone fossils in the Palaeozoic era. They rolled up when in danger, like their modern woodlice descendants.

TUFF—Rock formed of compacted volcanic ash.

UNCONFORMITY—A time break in the geological record, where rocks are not overlaid by the next group in the geological succession.

Bibliography and References

GENERAL GEOLOGICAL INTEREST

Fossils: a little guide in colour (Paul Hamlyn)
Himus, G. W. *Dictionary of Geology* (Penguin)
Himus, G. W., and Sweeting, M. M. *Elements of Field Geology* (Univ Tutorial Press)
Mesozoic Fossils (British Museum, Natural History)
Minerals: a little guide in colour (Paul Hamlyn)
Palaeozoic Fossils (British Museum, Natural History)

DEVONSHIRE GEOLOGY

Dines, H. G. *Metalliferous Mining Region of S. W. England*, vols 1 and 2 (HMSO 1956)
Directory of British Fossiliferous Localities (Palaeontographical Society 1954)
Exeter and Its Region (University of Exeter, for British Association meeting 1969)
Geological Survey Memoirs for the One-Inch maps
The Proceedings of the Ussher Society
Some Present Views of Aspects of the Geology of Devon and Cornwall commemorative volume for 1964. Royal Geol Soc Cornwall
South West England. British Regional Geology, 3rd edn (HMSO)

SHORT LIST OF REFERENCES AND USEFUL PAPERS

Arber, Muriel A. 'The Coastal Landslips of S E Devon', *Proc Geol Assoc*, 51 (1940), 257-71
Boylan, P. J. 'Dean William Buckland 1784-1856', *Studies in Speleology*, 1 no 5 (1967), 237-53

Bristow, C. M. 'A New Graphical Resistivity Technique for detecting air-filled cavities', *Studies in Speleology,* 1 no 4 (1966), 204-27

Brunsden, D. 'Denudation Chronology of the River Dart', *Trans Inst Brit Geogrs,* 32 (1963), 49-63

Brunsden, D. et al 'Denudation Chronology of Parts of S W England', *Field Studies,* 2 no 1 (1964), 115-32

Clarke, R. H. 'Quaternary Sediments off S E Devon', *Quart Jour Geol Soc,* 125 pt 3 (1970), 277-318

Dearman, W. R. 'Wrench Faulting in Cornwall and South Devon', *Proc Geol Assoc* 74 (1964), 265-87

Dineley, D. L. 'The Dartmouth Beds of Bigbury Bay, South Devon', *Quart Jour Geol Soc,* 122 (1966), 187-217

Dineley, D. L. 'The Devonian System in South Devonshire', *Field Studies,* no 1 (1961), 121-40

Durrance, E. M., and Hamblin, R. J. O. 'The Cretaceous Structure of Great Haldon, Devon', *Bull Geol Surv,* 30 (1969), 71-88

—— 'Ball Clay Production in S Devon', *Quarry Managers Journal* August 1964

Green, J. F. N. 'The History of the River Dart, Devon', *Proc Geol Assoc,* 60 (1949), 105-124

Johns, Ewart 'Langstone Rock—an experiment in the art of landscape description', *Geography,* 45 (1960), 176-82

Kidson, C. 'Dawlish Warren', *Trans Inst Brit Geogrs,* 16 (1950), 67-80

Knill, Diane C. 'The Permian Igneous Rocks of Devon', *Bull Geol Surv,* 29 (1969), 115-38

Laming, D. J. C. 'Age of the New Red Sandstone in South Devonshire', *Nature,* 207 (1965), 624-5

Laming, D. J. C. 'A Guide to the New Red Sandstone of Tor Bay, Petit Tor and Shaldon', *Trans Devon Assoc,* 101 (1969), 207-18

Laming, D. J. C. 'Imbrication, Palaeocurrents and Other Sedimentary Features in the Lower New Red Sandstone, Devonshire, England', *Jour Sedimentary Petrology,* (1966), 940-59

Orme, A. R. 'The Raised Beaches and Strandlines of South Devon', *Field Studies,* no 1 (1960), 109-30

Robinson, A. H. W. 'The Hydrography of Start Bay and its

relationship to beach changes at Hallsands', *Geog Jour* 127 pt 1 (1961), 63-77

Smith, W. E. 'The Cenomanian Deposits of S E Devonshire', *Proc Geol Assoc*, 72 (1961), 91-134

Smith, W. E. 'The Cenomanian Limestone of the Beer District, S Devon', *Proc Geol Assoc*, 68 (1957), 115-35

Smith, W. E. 'Summer Field Meeting in S Devon and Dorset', *Proc Geol Assoc*, 68 (1957), 136-52

Sutcliffe, A. J. 'Planning England's First Cave Studies Centre', *Studies in Speleology*, 1 no 2-3 (1965), 106-24

Taylor, P. W. 'The Plymouth Limestone', *Trans Roy Geol Soc Cornwall*, 18 (1950), 146-214

Vachell, E. T. 'Kents Cavern, Its Origin and History', *Trans Torquay Nat Hist Soc*, 11 pt 2

Walker, Miss H. H. 'Father John MacEnery, Scientist or Charlatan?', *Trans Devon Assoc*, 96 (1964), 125-46

Appendix

GENERALISED TABLE OF STRATA
MENTIONED IN THE TEXT

Era	*Period*	*Time*	*Duration*	*Strata*	
Quarternary	HOLOCENE	Present day		Beach, sand, earth, etc	
	PLEISTOCENE	1–1½ M	1 M	Clitters, head, cave earths, river gravels and terrace deposits	
Cainozoic	PLIOCENE	3–12 M	9 M		
	MIOCENE	12–25 M	13 M		
	OLIGOCENE	25–40 M	15 M	Bovey Clays	
	EOCENE	40–70 M	30 M	Flint Gravels of Haldon and Aller, Tower Wood Gravels	
Mesozoic	CRETACEOUS	70–135 M	65 M	Upper Chalk	
				Middle Chalk	
				Cenomanian limestone	
				Upper Greensand of Haldon and East Devon	
				Gault Clay	
	JURASSIC	135–180 M	45 M	Blue Lias	
				Rhaetic. White Lias	
	TRIASSIC	180–225 M	45 M	Tea Green Marls	NEW RED SANDSTONES
				Red Keuper Marls	
				Otter Sandstones	
				Budleigh Salterton Pebble Beds	
				Littleham Beds	
				Exmouth Beds	
Palaeozoic	PERMIAN	225–270 M	45 M	Exe Breccias	
				Dawlish Sands	
				Teignmouth Breccias	
	CARBONIFEROUS	270–350 M	80 M	Ness Beds and Oddicombe Breccias	
				Watcombe Breccias	
				Tor Bay Breccias and Sandstones	
			290 M	All existing beds folded and Dartmoor Granite intruded	
				Shales and sandstones	
	DEVONIAN	350–400 M	50 M	Upper Devonian of Saltern Cove	
				Middle Devonian Limestones	
				Middle Devonian Shales	
				Staddon Grits	LOWER DEVONIAN
				Meadfoot Beds	
				Dartmouth Slates	
				Start Point Schists—metamorphosed in Devonian times	

No rocks older than Devonian age are known in Devon. For each geological period, remember that this table is not a complete record since only those strata mentioned in this text are given.

Acknowledgements

I wish to express my sincere thanks to Dr E. B. Selwood for his great kindness in reading, advising and commenting on the text. Without his guidance on this and many previous occasions this book would not have been written. Also to Dr D. J. C. Laming for his help and advice on the New Red Sandstone material. He has given generously of his time.

Mr D. St John Thomas and Mrs M. Beckford gave me much initial encouragement. I am greatly indebted to Mr J. Sykes and the geologists of Messrs Watts, Blake, Bearne; to Mr E. T. Vachell for discussions on the Torquay area; to my wife for her understanding at all times and for reading and commenting on the script; and to Miss J. E. Cann for typing it.

I am also grateful to many other people who provided me with information I needed, answered my questions and helped in various other ways.

Figure 17 is after D. L. Dineley, and Figure 29 is based on section information supplied by E. T. Vachell.

Figure 42 is reproduced by permission of D. J. C. Laming and the Editor of the *Journal of Sedimentary Petrology.*

JOHN W. PERKINS

Index